T0258560

Principles and Practice of Fluid Mechanics

Principles and Practice of Fluid Mechanics

Edited by **Fay McGuire**

New York

Published by NY Research Press,
23 West, 55th Street, Suite 816,
New York, NY 10019, USA
www.nyresearchpress.com

Principles and Practice of Fluid Mechanics
Edited by Fay McGuire

International Standard Book Number: 978-1-63238-372-3 (Hardback)

Printed in the United States of America.

Contents

Preface

The world is advancing at a fast pace like never before. Therefore, the need is to keep up with the latest developments. This book was an idea that came to fruition when the specialists in the area realized the need to coordinate together and document essential themes in the subject. That's when I was requested to be the editor. Editing this book has been an honour as it brings together diverse authors researching on different streams of the field. The book collates essential materials contributed by veterans in the area which can be utilized by students and researchers alike.

Fluid mechanics is a complicated and extremely challenging field of study. It is central to several matters that are significant not only technologically, but also sociologically. This book emphasizes on a cross-section of procedures in fluid mechanics, each of which explains new ideas and relates to one or more matters of high interest during the early 21st century. The challenges constitute multi-phase flows, compressibility, non-linear dynamics, flow instability, transforming solid-fluid boundaries and fluids with solid-like characteristics. The procedures relate difficulties like weather and climate forecast, health care, efficiency of fuel, wind/wave energy harvesting, air quality, erosion, lessening landslides and noise.

Each chapter is a sole-standing publication that reflects each author's interpretation. Thus, the book displays a multi-facetted picture of our current understanding of application, resources and aspects of the field. I would like to thank the contributors of this book and my family for their endless support.

Editor

Part 1

General Fluid Mechanical Methods

Overview on Stereoscopic Particle Image Velocimetry

L. Martínez-Suástegui

ESIME Azcapotzalco, Instituto Politécnico Nacional, Colonia Santa Catarina,
Delegación Azcapotzalco, México, Distrito Federal,
Mexico

1. Introduction

Recently, Particle Image Velocimetry (PIV) has become the method of choice for multiple fluid-dynamic investigations (Adrian 1991; Willert and Gharib 1991; Jakobsen, Dewhirst et al. 1997; Westerweel 1997; Raffel, Willert et al. 1998). PIV is a reliable non-intrusive laser optical measurement technique that is based on seeding a flow field with micron-sized tracer particles and illuminating a two-dimensional (2D) slice or target area with a laser light sheet (Adrian and Yao 1985; Melling 1997). The target area is captured onto the sensor array of a digital camera, which is able to capture each light pulse in separate image frames. After recording a sequence of two light pulses, velocity vectors are derived from small subsections (called interrogation areas) of the target area of the particle-seeded flow by measuring the distance travelled by particles in the flow within a known time interval. The interrogation areas from the two image frames are cross-correlated with each other, pixel by pixel, producing a signal peak that allows for an accurate measurement of the displacement, and thus the velocity. Finally, the instantaneous velocity vector map over the whole target area is obtained by repeating the cross-correlation for each interrogation area over the two image frames captured by the camera (Nishino, Kasagi et al. 1989; Keane and Adrian 1992; Mao, Halliwel et al. 1993; Westerweel, Dabiri et al. 1997). The major drawback of the 2D PIV technique is that it records only the projection of the velocity vector into the plane illuminated by the laser sheet, so the out-of-plane velocity component is lost and the in-plane components are affected by an unrecoverable error due to the perspective transformation. Although the "classical" PIV method introduces an error due to the perspective projection and uncertainty in measuring the in-plane velocity components, most of the time it is commonly neglected, since it still allows the user to interpret the instantaneous flow field and its structures (Lawson and Wu 1997; Lawson and Wu 1997; Soloff, Adrian et al. 1997). Unfortunately, this is not the case when studying highly three-dimensional flows, where the only way to avoid the uncertainty error is to measure all three components of the velocity vectors using stereoscopic techniques (Adrian 1991; Arroyo and Greated 1991; Westerweel and Nieuwstadt 1991; Hinsch 1993; Prasad and Adrian 1993; Raffel, Gharib et al. 1995; Lawson and Wu 1999; Prasad 2000; Doorne 2004; Tatum, Finnis et al. 2005; Mullin and Dahm 2006; Tatum, Finnis et al. 2007). Stereoscopic PIV uses two cameras with separate viewing angles. By combining the two velocity fields measured by each camera using geometrical equations derived from the camera setup and a complicated calibration step, the third velocity component is evaluated and a three-dimensional (3D) velocity field is achieved (Nishino, Kasagi et al. 1989; Grant, Zhao et al. 1991; Raffel, Gharib et al. 1995; Willert 1997; Kähler and Kompenhans 2000; Shroder and Kompenhans 2004; Mullin and Dahm 2005; Perret, Braud et al. 2006).

In the present chapter, the technical basis, the set-up and components of a stereoscopic PIV apparatus/equipment are described so that the reader can understand how the different items of equipment are combined to form a coherent PIV tool (Hu, Saga et al. 2001). Afterwards, the principles of stereo PIV calibration, data acquisition, processing, and analysis are addressed. The aim of this chapter is to present in a more general context the aspects of the PIV technique relevant for those who intend to purchase a stereoscopic PIV system or those who want to perform stereoscopic PIV measurements. By understanding how to plan and perform experiments, it is hoped that it will allow the reader to successfully design a custom measurement system to fit a specific scientific or industrial application. In addition, this chapter will prove a valuable tool for those who already own a 2D PIV system and want to upgrade it for 3D measurement acquisition.

2. PIV system overview

There are several commercial stereoscopic PIV systems available, but the basic elements of these systems must include the following items: a pulsed laser system, two cameras for stereoscopic measurements, and a PC connected to a data acquisition card that synchronizes all of these items. The instrumentation required to perform the PIV data acquisition process is seeding, illuminating, recording, processing, and analysing the flow field. One drawback of PIV systems is that all of these items contribute to each stage of the measurement process, and therefore none of them can be spared. A brief description of the aforementioned instrumentation is presented in the following subsections.

2.1 Illumination systems

A stroboscopic light-sheet is desired to illuminate the plane of interest. This can be generated with pulsed lasers, continuous wave lasers, electro-optical shutters, polygon scanners, light guides and optical assemblies. Since only pulsed lasers have sufficient energy to record particle images, for relatively high speed flows seeded with micron or submicron particles, the most common choice are Nd:Yag lasers with a wavelength of 532 nm, since they offer repetition rates that match most of the commercially available CCD cameras. Pulsed laser systems include an array of optics with several cylindrical lenses that produces a diverging light sheet with adjustable thickness. This optical system includes and optical mount that can rotate through 360°. One thing to remember when designing experiments is that in order to avoid damage to the equipment, the laser must always be mounted and operated in a horizontal position. The sheet can be easily oriented by deflecting the laser beam using a mirror. Also, when it comes to choosing a pulsed laser for a stereo PIV system, the most important specifications to account for are: minimum sheet thickness range, the sheet focusing range, the maximum input pulse energy, the maximum input beam diameter, and of course its dimensions. The laser of choice must suit the measurement of the particular flow field under investigation. Nevertheless, when purchasing a PIV system, always aim for a laser with the maximum laser power output.

2.2 Cameras

Several CCD-based cameras are currently available. They differ in the desired spatial resolution, temporal resolution, directional ambiguity resolution, and cross-correlation options. Again, the cameras of choice depend on the resolution of the spatial and temporal features of the flow field to be measured. Generally, as the spatial resolution of the camera increases, its temporal resolution decreases. Therefore, when customizing your PIV equipment, always make sure that the chosen cameras meet the following criteria. They have enough temporal resolution in order to resolve the smallest velocity displacements between the first and second images of the particles of the flow field under investigation. The size of the smallest velocity structures can be measured in the flow field under study. One important thing to consider is that commercial PIV systems include a feature that interfaces the input buffer in the PIV processor with the cameras, thus allowing future upgrades of particular camera systems that better suit your needs. With that in mind, the best option when purchasing your first set of cameras for the PIV system is to make sure that they satisfy your particular needs in terms of temporal resolution with the highest spatial resolution.

2.3 Software to perform the PIV data acquisition process

The software of the PIV system must include a synchronization unit that links signals to and from the processor, the laser and cameras. Once the images have been acquired, the system must be able to produce and store vector maps or image maps in a database on a hard disc of a PC that keeps track of both the data and corresponding data acquisition and analysis parameters used.

2.4 3D traverse system

If the light sheet optics and the cameras are mounted on a common traverse system, they can be positioned at any desired point in the flow domain. This capability is particularly advantageous for measuring image data at multiple planes in a flow. Volume mapping is a technique based on performing multiple 3D stereoscopic PIV mappings in cross-sections of a flow within a very short time interval (Meinhart, Wereley et al. 2000; Klank, Goranovic et al. 2001). Unfortunately, this technique can only be used with a traverse system, and it is achieved by mounting the laser cavity and the cameras on an electronically controlled traverse system. In this way, when the entire traverse system is moved, the distance between the cameras and the light sheet remains constant so that there is no need to calibrate again. Although 2D or 3D PIV measurements can be performed without a traverse system, the main advantage of these systems is that they allow for fast and accurate calibration. For stereoscopic measurements in an enclosed flow (e.g., a duct flow, where the laser and cameras are outside of a transparent model), the index of refraction can have a strong effect on the calibration. One method to ensure that the two cameras have an orthogonal orientation with respect to the liquid-air interface is to redesign the wall of the test section to incorporate a triangular prismatic section. This is easily achieved by constructing a glass container that is filled with the same liquid and that is attached to the test section. By using a liquid prism between the test section and the lens of each camera, orthogonal viewing is accomplished with respect to the liquid-air interface and the aberrations are minimized (Prasad and Adrian 1993; Prasad and Jensen 1995). Also, when the test section has curved walls, distortion caused by refraction is minimized by enclosing the test section in a container with flat windows and filled with the same fluid as the test section.

3. Stereoscopic PIV calibration tools

In this section, the calibration tools for stereoscopic calibration and the calibration procedure are presented and described in detail.

3.1 Camera mounts for stereoscopic viewing

In most 3D PIV systems, when viewing the light sheet at an angle, the camera's entire field of view must be accurately focused. This is known as the Scheimpflug condition, and it is accomplished using camera mounts that include angle adjustment so that the image plane (CCD-plane), lens plane and object plane for each of the cameras intersect at a common point, as shown in Figure 1a (Scheimpflug 1904; Prasad and Jensen 1995; Zang and Prasad 1997). Figure 1b shows a camera mounted on a stereoscopic camera mount with angle adjustment and Scheimpflug condition.

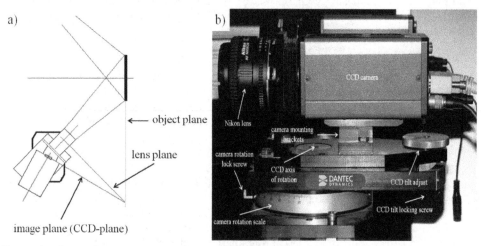

Fig. 1. a) Scheimpflug camera. b) CCD camera mounted on a Scheimpflug camera mount.

3.2 Calibration target

Stereo PIV measures displacements by using two cameras playing the role of eyes. By comparing the images of each camera against a calibration target, a stereoscopic calibration, which will be described in the next subsection, is achieved. Plane calibration targets consist of a one-sided white image with black dots on a regular spaced grid that is easily detected using image processing techniques (Harrison, Lawson et al. 2001; Ehrenfried 2002; Wieneke 2005). When these targets are used, the two cameras have to be positioned on the same side. On the other hand, double-sided (multi-level) targets contain a two-level grid of white dots on a black background located at two different and known orientations of the z-axis, where z is the distance away from the camera(s). One major advantage of these targets is that depending on the configuration of the experimental setup, the user can choose to mount the cameras either on opposite sides of the calibration target or on the same side. Although small angles between the two cameras can be used, the out-of-plane displacement is obtained more accurately when a larger angle is used between the two cameras. In this sense, the most accurate calibration is obtained when the angle between the two cameras is set to 90° (Sinha 1988; Westerweel and Nieuwstadt 1997). Nonetheless, stereoscopic PIV is also possible using a nonsymmetric arrangement of the cameras as long as the viewing axes are not collinear. Both types of targets have a larger centre dot called the "zero marker" which corresponds to the level at which the big dot in the centre of the calibration target is placed. Figure 2 shows the recommended setup for the cameras depending on the type of target used. Note that when the cameras have a narrow depth of field, a smaller separation angle will be needed to allow a wider field of view (i.e., where both cameras are in focus).

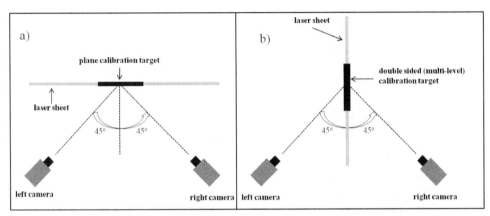

Fig. 2. Optimal camera configurations for an optimal determination of the stereoscopic calibration coefficients: a) Configuration using a plane (one-sided) calibration target. b) Configuration using a multi-level (two-sided) calibration target.

3.3 Calibration procedure

An imaging model describes the mapping of points from the image plane to object space, and the model parameters are determined through analysis of one or more calibration images. To obtain an imaging model, the first step is to accurately align the calibration target with the laser light sheet (Willert 1997). For 3D PIV measurements, the laser light thickness is adjusted by an optical arrangement supplied with the laser so that the illuminated plane is as thick as possible. If calibration is to be performed using a plane target, the cameras have to acquire images of the target through a number of z positions. This is generally accomplished by mounting the target on a special traverse unit and recording three to five z positions. Figure 3 shows the calibration grid images obtained by each camera for one z position using a plane calibration target of 100 x 100 mm with black dots and white background. The camera configuration corresponds to the one shown in Figure 2a).

Multi-level double sided targets eliminate the need for traversing, since they contain a two-level grid of white dots on a black background with known dot spacing in the x,y and z positions. Figure 4 shows a multi-level target of 270 x 190 mm with white dots and black background. The alignment of the laser light sheet and the target depend on the camera configuration. If the cameras are located on opposite sides of a multi-level calibration target, then the laser light sheet has to be aligned with the centre of the target. Here, the $z=0$ coordinate is located at the centre of the laser light sheet. If the cameras are placed on the same side of the calibration target, the laser light sheet has to be positioned in the plane located in the middle of each level. In this case, the plane located at the centre of the light sheet corresponds to the plane at which $z=0$. After aligning the calibration target, the user has to select the type of target used from a list. This setting allows the system to associate known x and y positions with the size and location of the markers, as the calibration algorithm automatically identifies the x and y coordinates on the images. The calibration target used during 3D stereo PIV measurements must always match the experimental setup, i.e. when measuring a large or small area, a large or small target is needed, respectively.

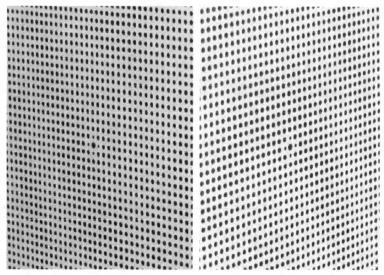

Fig. 3. Right and left calibration grid images obtained by each camera using a plane (one-sided) calibration target. The big dot in the centre of the calibration target is the zero marker.

Fig. 4. Multi-level target of 270 x 190 mm with white dots and black background.

Once the type of target used for the calibration is entered, information such as dot spacing in the x, y coordinates, zero marker diameter, axis marker diameter and level distance are known. The next step is to define the coordinate axes as seen from the cameras' point of view and the target configuration. Here, the x, y and z axes are always horizontal, vertical, and normal to the light sheet, respectively. However, each can be positive in any direction so that eight different coordinate system combinations are available. To obtain the z calibration coefficient, each camera records images of the calibration target for several z positions by traversing it in a direction normal to the laser sheet. Each displacement value is entered by the user in the dataset properties, and this calibration stage determines the x-y image to object plane space mapping (Lawson and Wu 1997; Soloff, Adrian et al. 1997). Table 1 shows some of the available sizes for each type of target. Note that for the case of multi-level targets, the z-coordinate refers to the location of the zero marker and the z values entered correspond to the dots located on the second level. These values are added or subtracted depending on the level spacing of each target and on how the calibration target is aligned with the laser light sheet.

Type of calibration target	Size (mm)	Z values entered (mm)
Dots	100 x 100	-
Dots	200 x 200	-
Dots	270 x 200	-
Dots	450 x 450	-
Multi-level	270 x 190	2nd level +4
Multi-level	270 x 190	2nd level -4
Multi-level	95 x 75	2nd level +2
Multi-level	95 x 75	2nd level -2

Table 1. Available sizes for plane and multi-level calibration targets.

After the entry properties of the calibration images are set and saved, the obtained transformation describes the overall perspective and lens distortion by providing parameter values for a specific image acquisition setup called the "imaging model fit". The calibration result can be verified by superimposing the model fit map to the corresponding calibration image. Figure 5 shows the imaging model for each camera based on an image size of 1344 x 1024 pixels after applying a direct linear transform and using a multi-level calibration target of 270 x 190 mm. The origin of the x, y and z coordinates is located at the centre of the zero marker and the coordinate axes are displayed as seen from each camera. The yellow dots at the centre of the white markers appear when a successful image model fit is obtained. Note that for both images, the Scheimpflug condition was not accomplished on the markers located at each corner. Nonetheless, a successful 3D calibration can still be obtained using an image model fit using sixteen out of the twenty available markers.

The quality of the calculated imaging model fit can be evaluated with the value of the average reprojection error. The latter describes the average pixel distance from every marker found to the predicted image location (distance from the yellow dots at the centre of the white markers to the points of intersection of the green grid in Figure 5). In this sense, the accuracy of the image model fit increases as the value of the average reprojection error decreases, and normally accepted values lie below 0.5. After performing a successful 3D calibration, the calibration target is removed and stereo measurements can be acquired by processing 2D PIV simultaneous recordings from each camera using the scale factor based on the image model fit. The two 2D vector maps recorded by each camera are then post processed using standard PIV processing software to obtain 3D vector maps. The main advantage of 3D calibration is that since a direct mapping function is derived between an object in 3D space and its corresponding location in the in-planes, there is no need to provide information regarding the geometric parameters of the stereoscopic image acquisition (Prasad 2000).

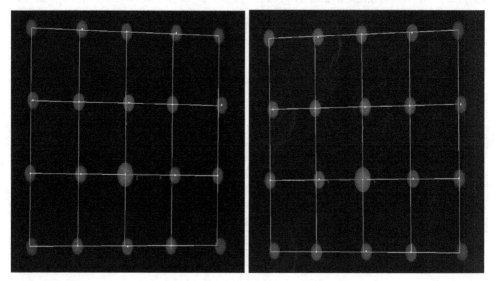

Fig. 5. Imaging model for each camera. The average reprojection errors are of 1.5544×10^{-1} and 2.2423×10^{-1} pixels for the left and right camera, respectively.

4. Data processing and analysis

This section describes how to perform 3D data processing and analysis after data acquisition, and addresses various options to export the processed information of the 3D image maps for further investigation and processing to spreadsheet displays, ASCII files, MATLAB or Tecplot 360.

4.1 Adaptive correlation

The adaptive correlation analysis method calculates velocity vectors starting with an initial interrogation area of size m x n pixels. After the first iteration, vectors are recalculated using a smaller interrogation area, and this procedure is repeated until a final interrogation area is reached. The number of iterations is specified by the user and it's called "number of refinement steps." The size of the final interrogation area is also specified by the user, and its value is determined by entering values of the horizontal and vertical sides (in pixels) of the latter. For each direction, available sizes are of 8, 16, 32, 64, 128 and 254 pixels. The initial interrogation area size is obtained by multiplying the final interrogation area size times the number of refinement steps, e.g., when selecting 3 refinement steps using a final interrogation area of 16 x 16 pixels, the initial interrogation area size is of 128 x 128 pixels. To reduce the correlation anomalies, overlap between neighbour interrogation areas can be specified independently for the horizontal and vertical directions. Validation parameters for the adaptive correlation method are normally used to remove spurious vectors. One of these parameters is the "peak validation." Here, the user sets values for the minimum and maximum peak widths as well as the minimum peak height ratio between the first and second peak. The "local neighbourhood validation" rejects spurious vectors and replaces them using a linear interpolation method based on the surrounding vectors located at an area of m x n pixels set by the user. Note that spurious vectors are identified by the inputted value of the "acceptance factor" parameter, and for larger values of this parameter, less velocity vectors are spatially corrected. The "moving average validation" method is used to validate vector maps by comparing each vector with the average of other vectors in a defined neighbourhood. Vectors that deviate from specified criteria are replaced using the average of the surrounding vectors. Figure 6 illustrates how the calculated velocity vectors obtained after applying an adaptive correlation can vary depending on the values of the input parameters described above. The computed vectors are displayed in blue, while the green vectors correspond to those obtained after interpolation with the surrounding vectors. This figure exemplifies the importance of adequately setting the values of the interrogation areas and validation methods employed. The instantaneous velocity map shown in Figure 6 is the recorded instantaneous flow structure of a free falling rotary seed that's spinning at a stationary height inside a low-speed, vertical wind tunnel crafted for studying its flow and kinematics. Velocity measurements were performed using a Dantec Dynamics DSPIV system and the images were processed using Dantec Dynamics software (DynamicStudio version 3.0.69). Seeding was supplied from a smoke generator (Antari Z-1500II Fog Machine, Taiwan, ROC) placed at the tunnel intake, and seeding quantity was regulated by monitoring the output from the DSPIV system (particle size 1 μm). To elucidate the effects of the value of the input parameters, the recipe for the left and right adaptive correlations of Figure 6 are displayed on the left and right sides of Figure 7, respectively.

The adaptive correlation on the left of Figure 6 used the following parameters for the interrogation areas: final interrogation area size of 32 x 16 pixels in the horizontal and vertical directions, respectively, no overlap between neighbour interrogation areas, one refinement step, and an initial interrogation area size is of 64 x 32 pixels. The validation parameters used are: moving average validation with an acceptance factor of 0.15 with three iterations using a neighbourhood size of 3 x 3. The adaptive correlation on the right of Figure 6 used the following parameters for the interrogation areas: final interrogation area size of 32 x 32 pixels, 25% and 50% of horizontal and vertical overlap, respectively, and five refinement steps using an initial interrogation area size of 1024 x 1024 pixels. The validation parameters used are: minimum peak height relative to peak 2 of 1.2, moving average validation using 3 iterations using a neighbourhood size of 3 x 3 and an acceptance factor of 0.15.

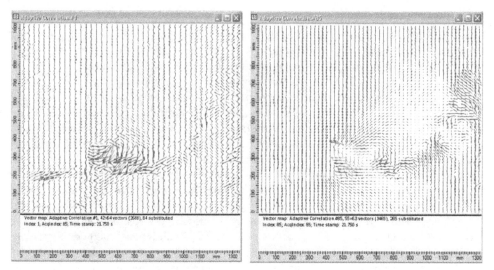

Fig. 6. Adaptive correlation applied to the same image pair by using different sizes of the final interrogation areas, number of refinement steps, and validation parameters.
Left image: 42 x 64 vectors, 2688 total vectors and 84 substituted vectors (green vectors).
Right image: 55 x 63 vectors, 3465 total vectors and 265 substituted vectors (green vectors).

Fig. 7. Adaptive correlation recipe used for the left image in Figure 6 (top and bottom left images), and adaptive correlation recipe used for the right image in Figure 6 (top and bottom right images).

4.2 Average filter

This technique is used to filter and smooth vector maps. To apply this method, the user defines the size of the $m \times n$ averaging area and an average vector is obtained based on the size of the averaging area size. Figure 8 shows how, after applying an average filter to the calculated velocity vectors, the instantaneous flow structure is smoothed. Note that a coherent structure is clearly visible at the centre of the right image. The recirculation shown corresponds to a leading-edge vortex (LEV) located on top of an autorotating airfoil.

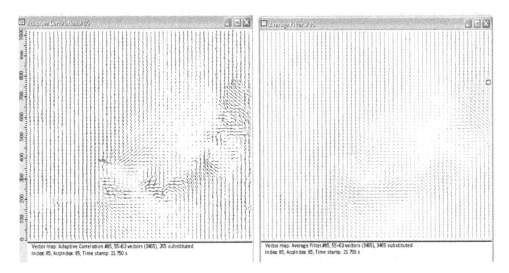

Fig. 8. Filtered instantaneous velocity map (right) after applying an average filter to the calculated velocity vectors (left) obtained using and adaptive correlation. The averaging area used is of 7 x 7 pixels.

4.3 Stereo PIV processing

As previously mentioned, the stereo PIV vector processing method computes 3D vectors based on the Imaging Model Fit obtained after stereoscopic calibration. To compute 3D vector maps, these steps must be followed: select the Image Model Fit and the 2D vector maps for each camera. Finally, apply the "Stereo Vector Processing" method from the PIV analysis group. Figure 9 illustrates how the resultant 3D PIV vector maps are layed out after applying the stereo vector processing method. Clearly, the flow field is displayed using traditional vector plots in 2D, and the scalar quantity of the out-of-plane velocity component is displayed with a scalar map and contours. Note that although the instantaneous 3D velocity field is obtained with this method, one major disadvantage is that the flow structure is still represented in the plane. Fortunately, commercial software from PIV systems has multiple options to export data to more powerful software packages for 3D flow visualization, such as MATLAB or Tecplot 360. Specifically, DynamicStudio from Dantec Dynamics includes a MATLAB link that transfers data of the recorded database to MATLAB's workspace. In addition, results obtained can be transferred back to the DynamicStudio database after processing data in MATLAB.

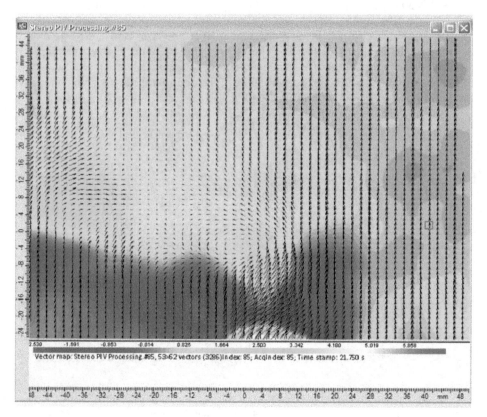

Fig. 9. 3D PIV vector map obtained after stereo PIV processing. The scale below the
vector map illustrates the magnitude and direction of the out-of-plane velocity component.

4.4 Scalar maps

Scalar maps are used to display on screen multiple data derived from the velocity fields.
Examples of scalar maps that can be calculated are: contours for the u, v and w velocity
components, contours with gradients of the u, v and w velocity components in the x, y and z
directions, vorticity contours, vortex identification methods, and the divergence of a 3D
vector field. Figure 10 shows the vorticity contours for the instantaneous flow structure
shown in Fig 9. In the next subsection, the steps to export databases for further processing
using Teclplot 360 are presented.

4.5 Exporting 3D vector maps for further processing using Tecplot 360

In this subsection, the steps to export data and reconstruct a three-component velocity field
using Tecplot 360 are described. The first step is to select the 3D PIV vector fields and scalar
maps to be exported. Afterwards, data are exported using a numerical export function. For
Tecplot 360, the user specifies the path to the directory where data will be saved, chooses the
names for the exported files, and saves them with a .DAT file extension. Finally, the

exported data files are loaded into Tecplot 360. Figure 11 displays the 3D instantaneous flow structure of Figure 9 and the vorticity contours of Figure 10 using Tecplot 360 after further processing. Figures 11a) and 11b) correspond to the frontal and rear perspectives normal to the plane of the laser sheet, respectively. The red circles enhance the location of the LEV and the trailing-edge vortex (TEV) close to the free falling rotary seed. Note how the LEV is visible using the front perspective, while the TEV is not. The opposite occurs for the rear perspective, where the TEV is visible but the LEV is not. Figure 12 shows the resulting flow structure after projecting the instantaneous 3D vector field of Figure 11 onto the plane illuminated by the laser sheet. Here, the vorticity contours are the same, but the size and location of the LEV, the TEV, and the overall flow structure changes dramatically. This exemplifies why 2D PIV measurements are prone to error when studying pronounced 3D flows, and hence, the importance of knowing the stereoscopic particle image velocimetry technique.

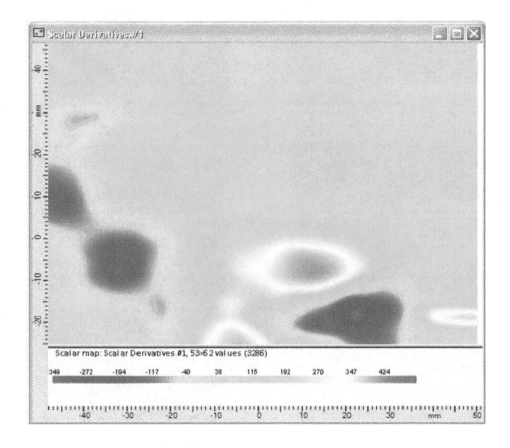

Fig. 10. Vorticity contours for the instantaneous flow structure shown in Figure 9.

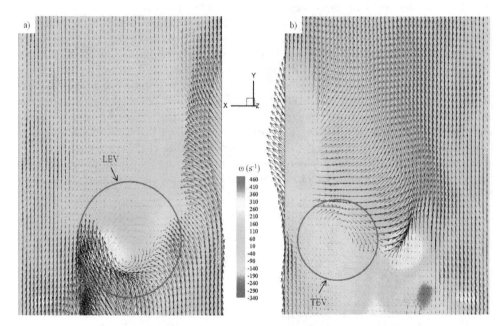

Fig. 11. Three dimensional instantaneous flow structure and vorticity contours displayed using Tecplot 360.

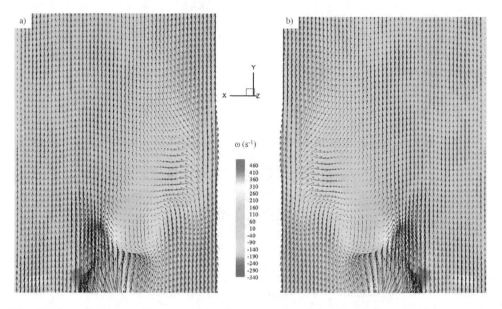

Fig. 12. Three dimensional instantaneous flow structure and vorticity contours projected onto a 2D plane and displayed using Tecplot 360.

5. Conclusion

Currently, the PIV technique is the standard method for measuring fluid flow velocity in a wide range of research and technology fields. Stereoscopic PIV is particularly well-suited for the study of biomedical flows (bifurcation flow phenomena in realistic lung models and heart valves), automotive industry (drag reduction, engine compartment flows and exhaust systems), aerospace industry (wind tunnel measurements for the study of wing design and trailing vortices), naval applications (propeller wake flow analysis), turbulent flow (jet mixing flows), combustion processes (fuel/air mixing, flame and fire research, jet propulsion, explosion research, exhaust control and swirl in combustion systems), oceanography (wave dynamics, sedimentation/particle transport, tidal modelling and river hydrology), design and optimization of electronic devices (thermal loading on electronic components), hydraulics and hydrodynamics (propulsion efficiency, pipe flows, channel flows and bubble dynamics), mixing processes, spray atomization, fluid structure interaction, vortex evolution and heat transfer studies. Nonetheless, one drawback of PIV is that it requires optical access for the light sheet as well as for the cameras, which may sometimes be difficult to ensure. In these cases, point based techniques such as pitot probes, Constant Temperature Anemometry (CTA), or numerical simulations are reliable tools for the measurement of flows. One major advantage is that 2D PIV systems can easily be expanded for 3D capabilities. Although commercial stereoscopic PIV systems equipped with a 3D traverse system are very expensive (at least US$300K), their ease of use allows anyone to perform high quality measurements without the need of any formal training in fluid mechanics. The aim of the present chapter is to describe the experimental methodology to plan and design experiments, perform a successful stereoscopic calibration, quantify the accuracy of the latter, and process the acquired data. It is hoped that the established methodology will prove useful to those who intent to obtain high quality three-component velocity results.

6. References

Adrian, R. J. (1991). "Particle-imaging techniques for experimental fluid mechanics." Annual Review in Fluid Mechanics 23: 261-304.

Adrian, R. J. and C. S. Yao (1985). "Pulsed laser technique application to liquid and gaseous flows and the scattering power of seed materials." Applied Optics 24: 44-52.

Arroyo, M. and C. Greated (1991). "Stereoscopic particle image velocimetry." Measurement Science and Technology 2: 1181-1186.

Doorne, C. W. H. v. (2004). Stereoscopic PIV on transition in pipe flow. The Netherlands, Delft University of Technology. Ph.D. thesis.

Ehrenfried, K. (2002). "Processing calibration grid images using the Hough transformation." Measurement Science and Technology 12: 975-983.

Grant, I., Y. Zhao, et al. (1991). Three component flow mapping: experiences in stereoscopic PIV and holographic velocimetry. New York: ASME.

Harrison, G. M., N. J. Lawson, et al. (2001). "The measurement of the flow around a sphere settling in a rectangular box using 3-dimensional particle image velocimetry." Chemical Engineering Communications 188: 143-178.

Hinsch, K. D. (1993). "Three-dimensional particle image velocimetry." Measurement Science and Technology 6: 742-753.

Hu, H. T. Saga, et al. (2001). "Dual-plane stereoscopic particle image velocimetry: system set-up and its application on a lobed jet mixing flow." Experiments in Fluids 31: 277-293.

Jakobsen, M. L., T. P. Dewhirst, et al. (1997). "Particle image velocimetry for predictions of acceleration fields and force within fluid flows." Measurement Science and Technology 8: 1502-1516.

Kähler, C. and J. Kompenhans (2000). "Fundamentals of multiple plane stereo particle image velocimetry." Experiments in Fluids 29: 70-77.

Keane, R. D. and R. J. Adrian (1992). "Theory of cross correlation analysis of PIV images." Journal of Applied Scientific Research 49: 191-125.

Klank, H., G. Goranovic, et al. (2001). "Micro PIV measurements in micro cell sorters and mixing structures with three-dimensional flow behaviour." Proceedings of 4th International Symposium on Particle Velocimetry. Göttingen, Institute of Aerodynamics and Flow Technology.

Lawson, N. and J. Wu (1997). "Three-dimensional particle image velocimetry: error analysis of stereoscopic techniques." Measurement Science and Technology 8: 894-900.

Lawson, N. J. and J. Wu (1997). "Three-dimensional particle image velocimetry: experimental error analysis of a digital angular stereoscopic system." Measurement Science and Technology 8: 1455-1464.

Lawson, N. J. and J. Wu (1999). "Three-dimensional particle image velocimetry: a low-cost 35mm angular stereoscopic system for liquid flows." Optics and Lasers in Engineering 32: 1-19.

Mao, Z. Q., N. A. Halliwell, et al. (1993). "Particle image velocimetry: high-speed transparency scanning and correlation-peak location in optical processing systems." Applied Optics 32(26): 5089-5091.

Meinhart, C. D., S. T. Wereley, et al. (2000). "Volume illumination for two-dimensional particle image velocimetry." Measurement Science and Technology 11: 809-814.

Melling, A. (1997). "Tracer particles and seeding for particle image velocimetry." Measurement Science and Technology 8: 1406-1416.

Mullin, J. A. and W. J. A. Dahm (2005). "Dual-plane stereo particle image velocimetry (DSPIV) for measuring velocity gradient fields at intermediate and small scales of turbulent flows." Experiments in Fluids 38: 185-196.

Mullin, J. A. and W. J. A. Dahm (2006). "Dual-plane stereo particle image velocimetry measurments of velocity gradient tensor fields in turbulent shear flow. I. Accuracy assessments." Physics of Fluids 18: 035101.

Nishino, N., N. Kasagi, et al. (1989). "Three dimensional particle image velocimetry based on automated digital image processing." Journal of Fluids Engineering 111: 384-391.

Perret, L., P. Braud, et al. (2006). "3-component acceleration field measurement by dual-time stereoscopic system." Experiments in Fluids 40: 813-824.

Prasad, A. (2000). "Stereoscopic particle image velocimetry." Experiments in Fluids 29: 103-116.

Prasad, A. K. and R. J. Adrian (1993). "Stereoscopic particle image velocimetry applied to liquid flows." Experiments in Fluids 15: 49-60.

Prasad, A. K. and K. Jensen (1995). "Scheimpflug stereocamera for particle image velocimetry to liquid flows." Applied Optics 34: 7092-7099.

Raffel, M., M. Gharib, et al. (1995). "Feasibility study of three-dimensional PIV by correlating images of particles within parallel light sheet planes." Experiments in Fluids 19(2): 69-77.

Raffel, M., C. Willert, et al. (1998). Particle Image Velocimetry A Practical Guide. Berlin, Springer.

Scheimpflug, T. (1904). Improved Method and Apparatus for the Systematic Alteration of Distortion of Plane Pictures and Images by Means of Lenses and Mirrors for Photography and for other purposes. B. P. N. 1196.

Schröder, A. and J. Kompenhans (2004). "Investigation of a turbulent spot using multi-plane stereo particle image velocimetry." Experiments in Fluids 36: 82-90.

Sinha, S. K. (1988). "Improving the accuracy and resolution of particle image or laser speckle velocimetry." Experiments in Fluids 6: 67-68.

Soloff, S., R. J. Adrian, et al. (1997). "Distortion compensation for generalized stereoscopic particle image velocimetry." Measurement Science and Technology 8: 1441-1454.

Tatum, J. A., M. V. Finnis, et al. (2005). "3-D particle image velocimetry of the flow field around a sphere sedimenting near a wall: Part 2. Effects of distance from the wall." Journal of Non-Newtonian Fluid Mechanics 127(2-3): 95-106.

Tatum, J. A., M. V. Finnis, et al. (2007). "3D particle image velocimetry of the flow field around a sphere sedimenting near a wall: Part 1. Effects of Weissenberg number." Journal of Non-Newtonian Fluid Mechanics 141: 99-115.

Westerweel, J. (1997). "Fundamentals of digital particle image velocimetry." Measurement Science and Technology 8: 1379-1392.

Westerweel, J., D. Dabiri, et al. (1997). "The effect of a discrete window offset on the accuracy of cross-correlation analysis of digital PIV recordings." Experiments in Fluids 23: 20-28.

Westerweel, J. and F. T. M. Nieuwstadt (1991). Performance tests on 3-dimensional velocity measurements with a two-camera digital particle-image velocimeter. ASME, New York, A. Dybbs and B. Ghorashi.

Westerweel, J. and F. T. M. Nieuwstadt (1997). Performance tests on 3-dimensional velocity measurements with a two-camera digital particle-image velocimeter. Proceedings of the 4th International Conference in Laser Anemometry - Advances and Applications, Cleveland, OH.

Wieneke, B. (2005). "Stereo-PIV using self-calibration on particle images." Experiments in Fluids 39: 267-280.

Willert, C. (1997). "Stereoscopic particle image velocimetry for applications in wind tunnel flows." Measurement Science and Technology 8: 1465-1479.

Willert, C. E. and M. Gharib (1991). "Digital particle image velocimetry." Experiments in Fluids 10: 181-193.

Zang, W. and A. K. Prasad (1997). "Performance evaluation of a Scheimpflug stereocamera for particle image velocimetry." Applied Optics 36(33): 8738-8744.

Part 2

Illustrative Applications

Two-Dimensional Supersonic Flow with Perpendicular Injection of the Gas

Ye. Belyayev[1] and A. Naimanova[2]

[1]*Al-Farabi Kazakh National University,*
[2]*Institute of Mathematics, Ministry of Education and Science,*
Kazakhstan

1. Introduction

The flow around jets has been comprehensively studied by many investigators. Mathematical models have been successfully realized for flow of perfect gases [1-4], but the practical design of supersonic ramjet (scramjet) engines requires an understanding of jet interaction with a crossflow for multispecies gases. The flow field in such devices is complicated by geometrical factors, turbulent fuel-air mixing, chemical reaction, shock waves, and separation regions ahead of and behind the jet. Despite successful numerical models of such flows, the detail of the flow physics of the combustion in the real devices through the use of computational tools is a difficult problem. Therefore the investigators studied some physical phenomena or proposed the numerical method which are important for solution of the problem of the combustion.

For example, this problem has been modeled by Drummond and Weidner [1-2] during that process Reynolds-averaged multispecies Navier-Stokes equations which were coupled with Baldwin-Lomax turbulence model have been solved by using the explicit predictor-corrector scheme of MacCormack. Drummond and Weidner modeled injection of fuel from two slots and studied influence of the slots width on the process of mixing. Their comparison of the pressure on the wall to the experiment data showed a satisfactory fit.

Grasso and Magi have also solved the Favre-averaged Navier-Stokes equations for a transverse gas injection in supersonic air streams [4]. The governing equations are solved by a finite volume approach with the $k-\varepsilon$ model equations coupled. A high-order TVD method has been developed to multispecies turbulent gas mixtures and the effect of the free stream Much number on the flow structure have been studied.

The numerical investigation of supersonic mixing of hydrogen with air has been performed in the work [5-6]. For the process the main flow air entering through a finite width of inlet and gaseous hydrogen is injected perpendicularly from the side wall. The explicit TVD scheme has been used to solve the system of two-dimensional Navier-Stokes equations. In this study the enhancement of mixing and good flame-holding capability of a supersonic combustor have been investigated by varying the distance of injector position from left boundary. Upstream recirculation can evolve which might be activated as a good flame holder.

One of the ways to solve Navier-Stokes equation for multispecies flow was proposed by Shuen and Yoon [7], during which they had developed numerical method using high-order lower-upper symmetric successive overrelaxation scheme. The higher approximation scheme for solution the Favre-averaged Navier- Stokes equations based on the ENO scheme had been constructed in [8-9]. The interaction of a planar supersonic air flow with perpendicularly injected hydrogen jet across the slot from the duct walls have been studied. The influence of the jet Mach number and the ratio of the jet and flow pressure on shock wave structure of the flow and the jet penetration depth was shown.

The $k-\varepsilon$ turbulence model developed by Jones and Launder had been successfully used to numerically calculate supersonic flows [10-11], where the effect of compressibility on the structure of turbulence must be considered. The compressibility correction to the dissipation-rate term in the k transport equation has been used by Rizzeta [10-11] to simulate a supersonic turbulent flowfield.

In this present work, a plane supersonic flow in a channel with perpendicular injection of jets from slots located symmetrically on the lower and upper walls is modeled numerically. For convenience of computations, symmetry is assumed so that only the lower half of the channel is modeled, with jet injection from the lower wall only. The flow pattern is shown in Figure 1.

The influence of the compressibility effect on the shock wave structure and on the vortex system ahead and behind of the jet are studied in the context of the $k-\varepsilon$ and algebraic Baldwin-Lomax models of turbulence. The detailed physical analysis of the supersonic jet interaction with a crossflow for multicomponent gases depending on parameters of the flowfield as the Mach numbers of the air stream, parameters of turbulence, pressure ratio are performed. Such an analysis can improve understanding of the relevant flow structures responsible for the generation of the pressure field and for the mixing of the injectant with the cross flow. Ultimately this analysis will contribute to understanding of scramjet fuel injection systems.

The choice of the range of parameters under consideration is determined by the available experimental data of the processes of hydrogen combustion in the range of regime parameters under consideration, which will allow comparisons with experimental data of other authors in the future.

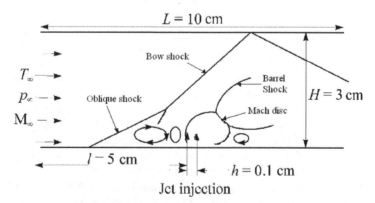

Fig. 1. Two dimensional slot injection flow scheme in the channel. L – length of channel, H – height of channel, l – length from the inlet to the slot, h – slot width.

2. Mathematical model

The two-dimensional Favre-averaged Navier-Stokes equations for multi-species flow with chemical reactions are:

$$\frac{\partial \vec{U}}{\partial t} + \frac{\partial (\vec{E} - \vec{E}_v)}{\partial x} + \frac{\partial (\vec{F} - \vec{F}_v)}{\partial z} = 0, \tag{1}$$

here the vector of the dependent variables and the vector of fluxes are given as

$$\vec{U} = \begin{pmatrix} \rho \\ \rho u \\ \rho w \\ E_t \\ \rho Y_k \\ \rho k \\ \rho \varepsilon \end{pmatrix} \quad \vec{E} = \begin{pmatrix} \rho u \\ \rho u^2 + p \\ \rho u w \\ (E_t + p)u \\ \rho u Y_k \\ \rho u k \\ \rho u \varepsilon \end{pmatrix} \quad \vec{F} = \begin{pmatrix} \rho w \\ \rho u w \\ \rho w^2 + p \\ (E_t + p)w \\ \rho w Y_k \\ \rho w k \\ \rho w \varepsilon \end{pmatrix}$$

$$\vec{E}_v = \left(0, \tau_{xx}, \tau_{xz}, u\tau_{xx} + w\tau_{xz} - q_x, J_{kx}, \frac{1}{\mathrm{Re}}\left(\mu_l + \frac{\mu_t}{\sigma_k}\right)\frac{\partial k}{\partial x}, \frac{1}{\mathrm{Re}}\left(\mu_l + \frac{\mu_t}{\sigma_\varepsilon}\right)\frac{\partial \varepsilon}{\partial x} \right)^T,$$

$$\vec{F}_v = \left(0, \tau_{xz}, \tau_{zz}, u\tau_{xz} + w\tau_{zz} - q_z, J_{kz}, \frac{1}{\mathrm{Re}}\left(\mu_l + \frac{\mu_t}{\sigma_k}\right)\frac{\partial k}{\partial z}, \frac{1}{\mathrm{Re}}\left(\mu_l + \frac{\mu_t}{\sigma_\varepsilon}\right)\frac{\partial \varepsilon}{\partial z} \right)^T,$$

The vector of source term is

$$\vec{W} = \left(0,0,0,0,0,\left[P_k - \rho\varepsilon\left(1 + \alpha M_t^2\right) + D \right], \left[C_{\varepsilon 1}P_k\varepsilon / k - C_{\varepsilon 2}f_{\varepsilon 2}\rho\varepsilon^2 / k + E_\varepsilon \right] \right)^T$$

The stress tensor components, heat and mass fluxes are:

$$\tau_{xx} = \frac{1}{\mathrm{Re}}\left(\mu_l + \frac{\mu_t}{\sigma_k}\right)\left(2u_x - \frac{2}{3}(u_x + w_z)\right); \quad \tau_{zz} = \frac{1}{\mathrm{Re}}\left(\mu_l + \frac{\mu_t}{\sigma_k}\right)\left(2w_z - \frac{2}{3}(u_x + w_z)\right);$$

$$\tau_{xz} = \tau_{zx} = \frac{1}{\mathrm{Re}}\left(\mu_l + \frac{\mu_t}{\sigma_k}\right)(u_z + w_x);$$

$$q_x = \frac{1}{\mathrm{Pr}\,\mathrm{Re}}\left(\mu_l + \frac{\mu_t}{\sigma_k}\right)\frac{\partial T}{\partial x} + \frac{1}{\gamma_\infty M_\infty^2}\sum_{k=1}^N h_k J_{xk}; \quad q_z = \frac{1}{\mathrm{Pr}\,\mathrm{Re}}\left(\mu_l + \frac{\mu_t}{\sigma_k}\right)\frac{\partial T}{\partial z} + \frac{1}{\gamma_\infty M_\infty^2}\sum_{k=1}^N h_k J_{zk};$$

$$J_{kx} = -\frac{1}{\mathrm{Sc}\,\mathrm{Re}}\left(\mu_l + \frac{\mu_t}{\sigma_k}\right)\frac{\partial Y_k}{\partial x}, \quad J_{kz} = -\frac{1}{\mathrm{Sc}\,\mathrm{Re}}\left(\mu_l + \frac{\mu_t}{\sigma_k}\right)\frac{\partial Y_k}{\partial z}.$$

Parameters of the turbulence are:

$$P_k = \tau_{txx}\frac{\partial u}{\partial x} + \tau_{txz}\frac{\partial u}{\partial z} + \tau_{tzx}\frac{\partial w}{\partial x} + \tau_{tzz}\frac{\partial w}{\partial z} \ ;$$

$$\tau_{txx} = \frac{\mu_t}{\text{Re}}\left(2\frac{\partial u}{\partial x} - \frac{2}{3}\left(\frac{\partial u}{\partial x} + \frac{\partial w}{\partial z}\right)\right); \quad \tau_{txz} = \frac{\mu_t}{\text{Re}}\left(\frac{\partial u}{\partial z} + \frac{\partial w}{\partial x}\right);$$

$$\tau_{tzz} = \frac{\mu_t}{\text{Re}}\left(2\frac{\partial w}{\partial z} - \frac{2}{3}\left(\frac{\partial u}{\partial x} + \frac{\partial w}{\partial z}\right)\right);$$

$$M_t^2 = 2M_\infty^2\,k\,/\,T \ ;$$

$$D = -2\cdot\frac{\mu_l}{\text{Re}}\left[\left(\frac{\partial k^{1/2}}{\partial x}\right)^2 + \left(\frac{\partial k^{1/2}}{\partial z}\right)^2\right];$$

$$E_\varepsilon = -\frac{2\mu_l\cdot\mu_t}{\rho\,\text{Re}^2}\left[\left(\frac{\partial^2 u}{\partial x^2}\right)^2 + \left(\frac{\partial^2 u}{\partial z^2}\right)^2 + \left(\frac{\partial^2 w}{\partial x^2}\right)^2 + \left(\frac{\partial^2 w}{\partial z^2}\right)^2\right];$$

$$f_{\varepsilon 2} = 1 - 0.3\exp\left(-\text{Re}_t^2\right); \quad \text{Re}_t = \text{Re}\left(\frac{\rho k^2}{\mu_l \varepsilon}\right);$$

$$C_{\varepsilon 1} = 1.44; C_{\varepsilon 2} = 1.92; \sigma_k = 1.0; \sigma_\varepsilon = 1.3 \ ,$$

where k,ε are the turbulent kinetic energy and rate of dissipation of turbulent kinetic energy respectively. P_k is turbulence production term, M_t is the turbulence Mach number, D, E_ε are the low-Reynolds number terms, and Y_k is the mass fraction of kth species, $k = 1...N$, where N is the number a component of a mix of gases.
The thermal equation for multi-species gas is:

$$p = \frac{\rho T}{\gamma_\infty M_\infty^2 W}, \quad W = \left(\sum_{k=1}^{N}\frac{Y_k}{W_k}\right)^{-1}, \quad \sum_{k=1}^{N}Y_k = 1, \tag{2}$$

where W_k is the molecular weight of the species.
The total energy equation is

$$E_t = \frac{\rho h}{\gamma_\infty M_\infty^2} - p + \frac{1}{2}\rho\left(u^2 + w^2\right) \tag{3}$$

The enthalpy of the gas mixture is calculated according to

$$h = \sum_{k=1}^{N}Y_k h_k \ , \quad h_k = h_k^0 + \int_{T_0}^{T}c_{pk}dT$$

where h is the specific enthalpy of kth species.
Specific heat at constant pressure for each component c_{pk} is:

$$c_{pk} = C_{pk} / W, \quad C_{pk} = \sum_{i=1}^{5} \bar{a}_{ki} T^{(i-1)}, \quad \bar{a}_{jk} = a_{jk} T_{\infty}^{j-1},$$

molar specific heat C_{pk} is given in terms of the fourth degree polynomial with respect to temperature in the JANAF Thermochemical Tables [12].
The system of equations (1) is written in the conservative, dimensionless form. The governing parameters are the entrance parameters at $x = -\infty$, which are the pressure and total energy normalized to $\rho_{\infty} u_{\infty}^2$, the enthalpy normalized to $R_0 T_{\infty} / W_{\infty}$, the molar specific heat normalized to R_0, the turbulence kinetic energy k normalized to u_{∞}^2, the dissipation of turbulence kinetic energy ε normalized to u_{∞}^3 / h and the considered lengths normalized to the slot width h.

3. Transport properties

The coefficient of dynamic viscosity is represented in the form of the sum of μ_l - molecular viscosity and μ_t - turbulent viscosity:

$$\mu = \mu_l + \mu_t \tag{4}$$

The Wilke formula is used to determine the mixture viscosity coefficient by

$$\mu_l = \sum_{i=1}^{N} \frac{X_i \mu_i}{\Phi_i},$$

where

$$\Phi_i = \sum_{r=1}^{N} X_r \left[1 + \sqrt{\frac{\mu_i}{\mu_r}} \left(\frac{W_r}{W_i} \right)^{1/4} \right]^2 \left[\sqrt{8} \sqrt{1 + \frac{W_i}{W_r}} \right]^{-1}$$

μ_i is the molecular viscosity of species i, which is defined by following formula [13, p.16]

$$\mu_i = \frac{\mu_{i\infty}}{\mu_{\Lambda\infty}} \sqrt{W_i T_{\infty}}$$

where

$$\mu_{i\infty} = 2.6693 \times 10^{-7} \frac{\sqrt{W_{i\infty} T_{\infty}}}{\sigma_i^2 \Omega_i^{(2.2)*} (T_i^*)}, \quad \mu_{\Lambda\infty} = \sum_{i=1}^{N} \frac{X_i \mu_{i\infty}}{\Phi_i} \tag{5}$$

In (5), σ_i is the collision diameter of species i :

$$\sigma_1 = 2.63; \quad \sigma_2 = 3.30; \quad \sigma_3 = 3.5; \quad \sigma_4 = 3.050; \quad \sigma_5 = 0.50; \quad \sigma_6 = 0.560; \quad \sigma_7 = 3.50,$$

where $\Omega_i^{(2.2)*}$ is the collision integral, T_i^* - kT / ε_i is the reduced temperature, and $\frac{\varepsilon_i}{k}$ is the potential. According to [13, p.17], $\Omega_i^{(2.2)*}(T_i^*) = 1$.

Turbulent viscosity μ_t is defined by using a $k - \varepsilon$ turbulence model

$$\mu_t = C_\mu f_\mu \operatorname{Re}_t \mu_l, \text{ where } C_\mu = 0.09 \tag{6}$$

$$Re_t = Re\left(\frac{\rho k^2}{\mu_l \varepsilon}\right) \quad f_\mu = \exp\left[\frac{-3.4}{(1 + 0.02 Re_t)^2}\right].$$

4. Initial and boundary conditions

At the entrance:

$$W_k = W_{k\infty}, \ p = p_\infty, \ T = T_\infty, \ u = M_\infty\sqrt{\frac{\gamma_\infty R_0 T_\infty}{W_\infty}}, \ w = 0, \ Y_k = Y_{k\infty}, \ x = 0, \ 0 \le z \le H.$$

k, ε at the entrance:

the k, ε profile at the entrance is defined by using Baldwin-Lomax's algebraic turbulence model for known averaged physical parameters of the flow, where the coefficient of turbulent viscosity contains local velocity gradients in the transverse direction. Making an assumption $P_k / \varepsilon = 1$ we find

$$k = k_\infty, \ k_\infty = \frac{\mu_t P_k}{\rho \operatorname{Re}\sqrt{C_\mu f_\mu}}, \text{ where } P_k = \frac{\mu_t}{\operatorname{Re}}\left(\left(\frac{\partial w}{\partial x}\right)^2 + \frac{4}{3}\left(\frac{\partial w}{\partial z}\right)^2\right)$$

$$\varepsilon = \varepsilon_\infty, \ \varepsilon_\infty = C_\mu f_\mu \operatorname{Re}\left(\frac{\rho k^2}{\mu_t}\right).$$

The boundary layer on the wall is given in the input section, and the velocity profile approximated by power law.

On the slot:

$$W_k = W_{k0}, \ p = np_\infty, \ T = T_0, \ w = M_0\sqrt{\frac{\gamma_0 R_0 T_0}{W_0}}, \ u = 0, \ Y_k = Y_{k0}, \ z = 0, \ L_b \le x \le L_b + h.$$

On the slot for $k - \varepsilon$,

$$k = k_0, \ k_0 = \frac{3w^2 T_i}{2},$$

where T_i is the jet turbulence intensity. Turbulence energy dissipation is defined by assumption $\mu_t = \mu_l$

$$\varepsilon = \varepsilon_0, \ \varepsilon_0 = \frac{\rho_0 k_0^2}{53.56 \mu_l},$$

($n = p_0 / p_\infty$ is the jet pressure ratio, p_0 is the jet pressure, and p_∞ is the flow pressure). On the lower wall the no-slip condition is imposed, for the temperature the adiabatic condition are imposed; on the upper boundary the condition of symmetry is assumed; on the outflow the nonreflecting boundary condition [9] is used.

5. Method of the solution

To take into account the flow in the boundary layer, near the wall, and near the slot, i.e., in regions of high gradients, more accurately, we refine the grid in the longitudinal and transverse directions by the transformations

$$\frac{\partial \tilde{U}}{\partial t} + \frac{\partial \tilde{E}}{\partial \xi} + \frac{\partial \tilde{F}}{\partial \eta} = \frac{\partial \tilde{E}_{v2}}{\partial \xi} + \frac{\partial \tilde{E}_{vm}}{\partial \xi} + \frac{\partial \tilde{F}_{v2}}{\partial \eta} + \frac{\partial \tilde{F}_{vm}}{\partial \eta}, \tag{7}$$

where

$$\tilde{U} = \bar{U}/J, \ \tilde{E} = \xi_x \bar{E}/J, \ \tilde{F} = \eta_z \bar{F}/J, \ \tilde{E}_{v2} = \xi_x \bar{E}_{v2}/J, \ \tilde{E}_{vm} = \xi_x \bar{E}_{vm}/J, \ \tilde{F}_{v2} = \eta_z \bar{F}_{v2}/J,$$
$$\tilde{F}_{vm} = \eta_z \bar{F}_{vm}/J, \text{ and } J = \partial(\xi,\eta)/\partial(x,z)$$

is the Jacobian of mapping..
System (7) linearized with respect to the vector U in form:

$$\tilde{U}^{n+1} + \Delta t \left(\frac{\partial \tilde{E}^{n+1}}{\partial \xi} + \frac{\partial \tilde{F}^{n+1}}{\partial \eta} - \frac{\partial \tilde{E}_{vm}^{n+1}}{\partial \xi} - \frac{\partial \tilde{E}_{v2}^{n+1}}{\partial \xi} - \frac{\partial \tilde{F}_{vm}^{n+1}}{\partial \eta} - \frac{\partial \tilde{F}_{v2}^{n+1}}{\partial \eta} \right) = \tilde{U}^n + O(\Delta t^2). \tag{8}$$

Here,

$$\tilde{E}^{n+1} \approx A_\xi^n \tilde{U}^{n+1}, \quad \tilde{F}^{n+1} \approx B_\eta^n \tilde{U}^{n+1}, \tag{9}$$

$$A_\xi = \xi_x A, \quad B_\eta = \eta_z B,$$

where $A = \partial \bar{E} / \partial \bar{U}$ and $B = \partial \bar{F} / \partial \bar{U}$ are the Jacobian matrices [8-9].
The terms containing the second derivatives are presented as sums of two vectors:

$$\tilde{E}_{v2}^{n+1} = \tilde{E}_{v21}^{n+1} + \tilde{E}_{v22}^n, \quad \tilde{F}_{v2}^{n+1} = \tilde{F}_{v21}^{n+1} + \tilde{F}_{v22}^n, \tag{10}$$

where the vectors $\tilde{E}_{v21}^{n+1}, \tilde{F}_{v21}^{n+1}$ are written in the following form:

$$\tilde{E}_{v11}^{n+1} = \frac{\mu_t \xi_x}{ReJ} \left[0, \frac{4}{3} \frac{\partial}{\partial \xi} \left(\frac{u \ \rho}{\rho} \right)^{n+1}, \frac{\partial}{\partial \xi} \left(\frac{w \ \rho}{\rho} \right)^{n+1}, \frac{\gamma}{Pr} \frac{\partial}{\partial \xi} \left(\frac{E_t}{\rho} \right)^{n+1} \right]^T,$$

$$\tilde{F}_{v21}^{n+1} = \frac{\mu_t \eta_z}{ReJ} \left[0, \frac{\partial}{\partial \eta} \left(\frac{u \ \rho}{\rho} \right)^{n+1}, \frac{4}{3} \frac{\partial}{\partial \eta} \left(\frac{w \ \rho}{\rho} \right)^{n+1}, \frac{\gamma}{Pr} \frac{\partial}{\partial \eta} \left(\frac{E_t}{\rho} \right)^{n+1} \right]^T,$$

and the vectors $\tilde{E}_{v12}^n$, $\tilde{F}_{v22}^n$ contain the remaining dissipative functions of the form:

$$\tilde{E}_{v12}^n = \frac{\xi_x^2}{ReJ}\left[0,0,0,\left[\left(\mu_t - \frac{\gamma\mu_t}{Pr}\right)\left(w\frac{\partial w}{\partial \xi}\right) + \left(\frac{4}{3}\mu_t - \frac{\gamma\mu_t}{Pr}\right)u\frac{\partial u}{\partial \xi}\right]^n\right]^T,$$

$$\tilde{F}_{v22}^n = \frac{\eta_z^2}{ReJ}\left[0,0,0,\left[\left(\mu_t - \frac{\gamma\mu_t}{Pr}\right)\left(u\frac{\partial u}{\partial \eta}\right) + \left(\frac{4}{3}\mu_t - \frac{\gamma\mu_t}{Pr}\right)w\frac{\partial w}{\partial \eta}\right]^n\right]^T,$$

$$\tilde{E}_{vm} = \frac{\xi_x\mu_t}{ReJ}\left[0,-\frac{2}{3}\left(\eta_z\frac{\partial w}{\partial \eta}+\zeta_y\frac{\partial v}{\partial \zeta}\right),\eta_z\frac{\partial u}{\partial \eta},-\frac{2}{3}\left(\zeta_y u\frac{\partial v}{\partial \zeta}+\eta_z u\frac{\partial w}{\partial \eta}\right)+\right.$$
$$\left.+\left(\eta_z w\frac{\partial u}{\partial \eta}+\zeta_y v\frac{\partial u}{\partial \zeta}\right)\right],$$

$$\tilde{F}_{vm} = \frac{\eta_z\mu_t}{ReJ}\left[0,\eta_z\frac{\partial w}{\partial \eta},-\frac{2}{3}\left(\xi_x\frac{\partial u}{\partial \xi}+\zeta_y\frac{\partial w}{\partial \zeta}\right),\left(\xi_x u\frac{\partial w}{\partial \xi}+\zeta_y u\frac{\partial w}{\partial \zeta}\right)-\right.$$
$$\left.-\frac{2}{3}\left(\xi_x w\frac{\partial u}{\partial \xi}+\zeta_y w\frac{\partial v}{\partial \zeta}\right)\right].$$

According to a principle of construction ENO scheme the system (7) for integration on time is formally represented as:

$$\Delta\tilde{U}^{n+1} + \Delta t\left[\left(\hat{A}^+ + \hat{A}^-\right)\frac{\vec{E}^m}{\xi} + \left(\hat{B}^+ + \hat{B}^-\right)\frac{\vec{F}^m}{\eta} - \right.$$
$$\left.\left[\frac{(\tilde{E}_{v2}^{n+1} + \tilde{E}_{vm}^n)}{\xi} - \frac{(\tilde{F}_{v2}^{n+1} + \tilde{F}_{vm}^n)}{\eta}\right]\right] = O\left(\frac{1}{2}\Delta t^2\right)$$
(11)

Here $\vec{E}^m$, $\vec{F}^m$ is called the modified flux vector. It consists from the original flux vector ($\tilde{E}$, $\tilde{F}$ and additional terms of third-order accuracy $\vec{E}_\xi,\vec{D}_\xi$, $\vec{E}_\eta,\vec{D}_\eta$):

$$\vec{E}^m = \tilde{E}^{n+1} + (\vec{E}_\xi + \vec{D}_\xi)^n,$$
(12)

modified flux $\vec{F}^m$ is written similarity and $\hat{A}^+ + \hat{A}^- = I$, $\hat{A}^\pm = A^\pm A^{-1}$, $\hat{B}^\pm = B^\pm B^{-1}$, I - unity matrix.

Applying factorization to (11), we obtain two one-dimensional operators, which are resolved by matrix sweep:

Step 1.

$$\left[I+\Delta t\left\{(\hat{A}_{i-1/2}^+\Delta_-A_\xi^n + \hat{A}_{i+1/2}^-\Delta_+A_\xi^n) + \Delta\frac{\mu_t\xi_x^2}{ReJ}\Delta\frac{1}{U_1^n}\right\}\right]U^* = RHS_\xi^n + RHS_\eta^n$$

Step 2.

$$\left[I+\Delta t\left\{(\hat{B}_{j-1/2}^+\Delta_-B_\eta^n + \hat{B}_{j+1/2}^-\Delta_+B_\eta^n) + \Delta\frac{\mu_t\eta_z^2}{ReJ}\Delta\frac{1}{U_1^n}\right\}\right]\tilde{U}^{n+1} = U^*,$$
(13)

$$RHS_\xi^n = \hat{A}_{i+1/2j}^- \left[\left(\vec{E}_\xi + \vec{D}_\zeta \right)_{i+1j} - \left(\vec{E}_\xi + \vec{D}_\zeta \right)_{ij} \right]^n + \hat{A}_{i-1/2j}^+ \left[\left(\vec{E}_\xi + \vec{D}_\zeta \right)_{ij} - \left(\vec{E}_\xi + \vec{D}_\zeta \right)_{i-1j} \right]^n,$$

$$\hat{A}_{i+1/2j}^- \left[\left(\vec{E}_\zeta + \vec{D}_\zeta \right)_{ij} \right]^n = \begin{array}{l} \left(minmod \left(\overline{E}_{\xi i+1/2j} \; \overline{E}_{\xi i-1/2j} \right) + \right. \\[2mm] + \begin{cases} \dot{m} \left(\Delta_- \hat{D}_{\xi i+1/2j}, \; \Delta_+ \hat{D}_{\xi i+1/2j} \right) & \text{if } \left| \Delta_- \tilde{U}_{ij} \right| > \left| \Delta_+ \tilde{U}_{ij} \right| \\[2mm] \dot{m} \left(\Delta_- \overline{D}_{\xi i-1/2j}, \; \Delta_+ \overline{D}_{\xi i-1/2j} \right) & \text{if } \left| \Delta_- \tilde{U}_{ij} \right| \le \left| \Delta_+ \tilde{U}_{ij} \right| \end{cases} \end{array},$$

$$\hat{A}_{i-1/2j}^+ \left[\left(\vec{E}_\xi + \vec{D}_\zeta \right)_{ij} \right]^n = \begin{array}{l} R \, \hat{\Lambda}^+ R_{i-1/2j}^{-1} \left[\left(minmod \left(\overline{E}_{\xi i+1/2j} \; \overline{E}_{\xi i-1/2j} \right) - \right. \right. \\[2mm] - \begin{cases} \dot{m} \left(\Delta_- \hat{D}_{\xi i-1/2j}, \; \Delta_+ \hat{D}_{\xi i-1/2j} \right) & \text{if } \left| \Delta_- \tilde{U}_{ij} \right| \le \left| \Delta_+ \tilde{U}_{ij} \right| \\[2mm] \dot{m} \left(\Delta_- \overline{D}_{\xi i-1/2j}, \; \Delta_+ \overline{D}_{\xi i-1/2j} \right) & \text{if } \left| \Delta_- \tilde{U}_{ij} \right| \le \left| \Delta_+ \tilde{U}_{ij} \right| \end{cases} \end{array},$$

$$\overline{E}_{\xi i\pm1/2j} = \left(R \, sign(\Lambda) R^{-1} \right)_{i\pm1/2j} \frac{1}{2} \left[I - \frac{\Delta t}{\Delta \xi} \left(R \left| \Lambda \right| R^{-1} \right)_{i\pm1/2} \right] \Delta_\pm \tilde{E}_{ij},$$

$$\overline{D}_{\xi i\pm1/2j} = \left(R \, sign(\Lambda) R^{-1} \right)_{i\pm1/2j} \frac{1}{6} \left[\frac{\Delta t^2}{\Delta \xi^2} \left(R \left| \Lambda \right| R^{-1} \right)_\pm^2 - I \right] \Delta_\pm \tilde{E}_{ij},$$

$$\hat{D}_{\xi i\pm1/2j} = \overline{E}_{\xi i\pm1/2j} + \overline{D}_{\xi i\pm1/2j},$$

where

$$minmod(a,b) = \begin{cases} s \times min \left(\left| a \right|, \left| b \right| \right) & \text{if} & sign(a) = sign(b) = s \\ 0 & & other \end{cases},$$

$$\dot{m}(a,b) = \begin{cases} a & \text{if } \left| a \right| \le \left| b \right| \\ b & \text{if } \left| a \right| > \left| b \right| \end{cases}.$$

The second term RHS_η^n is written similarly.

In approximation of derivatives in convective and diffusion terms, we use second-order central-difference operators.

The numerical solution of the system (7) is calculated in two steps. The first determines the dynamic parameters and second determines the mass species.

Then it is necessary to define Jacobian matrix which, in a case of the thermally perfect gas, represents a difficult task. This problem is connected with the explicit representation of pressure through the unknown parameters. Here pressure is determined by introducing an effective adiabatic parameter of the gas mixture.

$$\overline{\gamma} = \frac{h_{sm}}{e_{sm}}, \tag{14}$$

where

$$h_{sm} = \sum_{i=1}^{N} Y_i \int_{T_0}^{T} c_{p_i} dT$$

and

$$e_{sm} = \sum_{i=1}^{N} Y_i \int_{T_0}^{T} c_{v_i} dT$$

are the enthalpy and internal energy of the mixture minus the heat and energy of formation; $T_0 = 293K$ is the standard temperature of formation, which allows to write an expression for the pressure

$$p = (\bar{\gamma} - 1)\left[E_t - \frac{1}{2}\rho\left(u^2 + w^2\right) - \rho\frac{h_0}{\gamma_\infty M_\infty^2}\right] + \frac{\rho T_0}{M_\infty^2 W}.$$

The temperature is found by using the Newton-Raphson iteration [8-9].

6. Numerical results

The parameters of coordinate transformation (7) have the form:

$$\xi(x) = K + \frac{1}{\tau} arsh\left[\left(\frac{x}{x_c} - 1\right)sh(\tau K)\right],$$

$$\eta(z) = H\left[(\beta + 1) - (\beta - 1)\left(\frac{\beta + 1}{\beta - 1}\right)^{1 - \frac{z}{a}}\right] \bigg/ \left[\left(\frac{\beta + 1}{\beta - 1}\right)^{1 - \frac{z}{a}} + 1\right],$$

$$K = \frac{1}{2\tau}\ln\left[\left(1 + (e^\tau - 1)\frac{x_c}{L}\right) \bigg/ \left(1 - (e^\tau - 1)\frac{x_c}{L}\right)\right],$$

β and τ are refinement factors ($\beta > 1$ and $\tau > 1$), a is the height of the computational domain in the generalized coordinates, and x_c is the point with respect to which grid refinement is performed.

The use of algorithm (13) with the Baldwin-Lomax turbulence model has been tested by authors and results were stated in reference [8].

Numerical simulation was performed for perpendicular injection of a gas into two-dimensional channel with the channel height of - 7,62 cm, length - 15 cm and width of slot 0,0559 cm. The slot was located at a distance of 10 cm from entrance. The parameters were $M_0 = 1$, $T_0 = 642$ K, $T_\infty = 800$ K. The pressure ratios ranged from 2.75 to 20, and the freestream Mach number M_∞ ranged from 1.75 to 4.75. The space grid was 241x181. The correction coefficient for compressibility was taken as $\alpha = 0.3$.

The Figure 2 shows the influence of turbulence model on the character of interaction of the jets with the supersonic air flow. In this figure, the shaded pictures of pressure calculated by Baldwin-Lomax and $k - \varepsilon$ turbulence models with and without compressibility correction are represented.

As is well known, the physics flow is the following: due to deceleration of the main flow, the pressure ahead of the jet increases, and a bow shock wave is formed. An oblique shock wave emanates upstream from the bow shock wave. A second shock wave, named the barrel shock emerged behind the oblique shock wave. The bow shock wave, oblique shock wave and barrel shock wave intersect at one point to form a complicated λ-shaped structure.

All of these shock waves are presented in the Figure 2 for different freestream Mach numbers. Although the qualitative picture of the shock waves are similar for both turbulence models, the evident divergence of the λ-shaped structure is observed in the turbulent boundary layer, which is more prominent for smaller M_∞ Figure 2 (a-c).

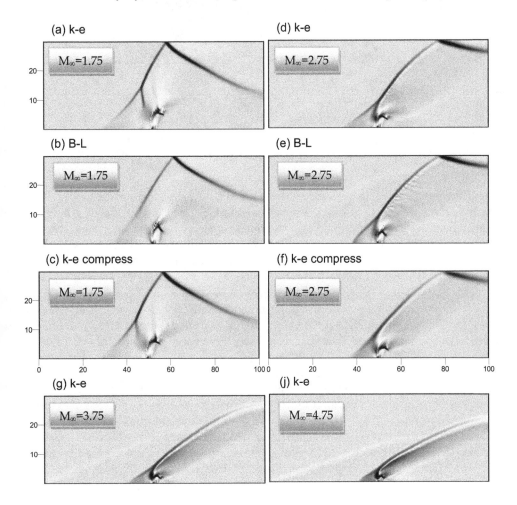

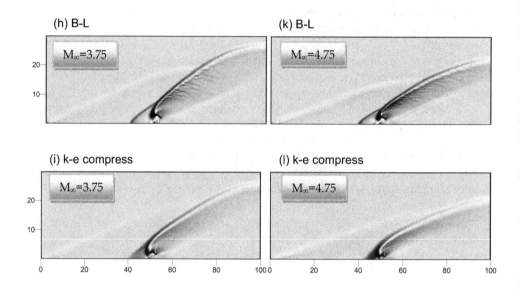

(a)–(c) $M_\infty = 1.75$, (d)–(f) $M_\infty = 2.75$
(g)–(i) $M_\infty = 3.75$, (j)–(l) $M_\infty = 4.75$

Fig. 2. Shaded picture of pressure for various Mach number with $k-\varepsilon$, B-L, and $k-\varepsilon$ compressibility turbulence models

The differences among the models in the boundary layer are also evident in the local Mach number contours. The distribution of the iso-Mach line contours is shown in Figure 3. In all cases the sonic velocity of the jet becomes supersonic at a certain distance because of the acceleration immediately after injection. Then, the boundary of the supersonic region is closed, forming a barrel which is a consequence of interaction of the shock waves system in the jet itself. With increasing freestream Mach number the jet penetration is reduced, due to the reduction of the hydrogen momentum with respect to the incoming air momentum and consequently a significant decrease in the barrel size is observed. The iso-Mach line contours in the subsonic region (Figure 3) were compared, and quantitative differences among the models can be seen. It is obvious that increasing of M_∞ reduces these differences.

The numerical experiments have shown that the acceleration of the flow inside of the barrel is larger for the $k-\varepsilon$ model than for the algebraic model. For example, in the case of $k-\varepsilon$ model the flow inside of the barrel for $M_\infty = 1.75$ is accelerated until the maximum local Mach number becomes $M = 3.4$ whereas in the algebraic model the local Mach number only reaches $M = 2.8$. Simulations that used different values of M_∞ revealed that the acceleration of both models is decreased with increaseed Mach number; for the case of $M_\infty = 4.75$ for $k-\varepsilon$ model maximum inside of barrel is $M = 2.87$ and for algebraic $M = 2.56$.

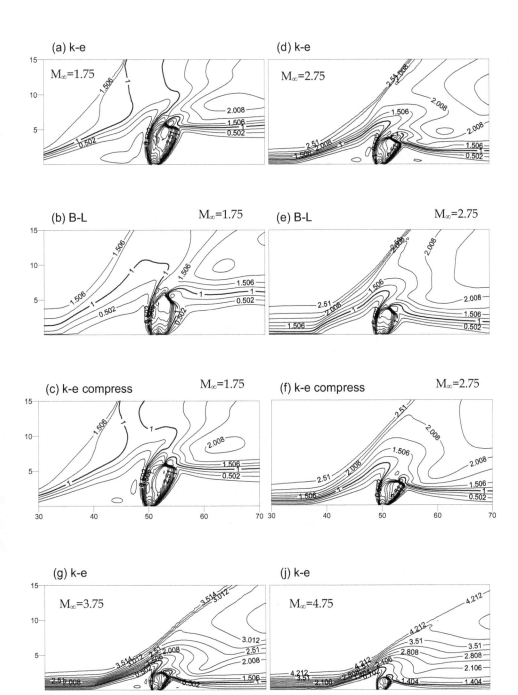

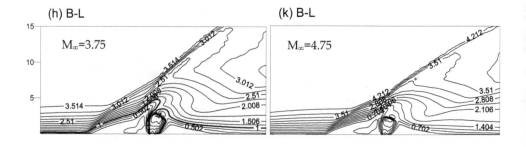

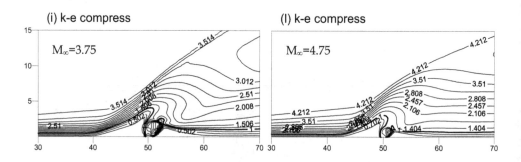

(a)–(c) $M_\infty = 1.75$, (d)–(f) $M_\infty = 2.75$
(g)–(i) $M_\infty = 3.75$, (j)–(l) $M_\infty = 4.75$

Fig. 3. iso-Mach line contours for various Mach number with $k-\varepsilon$, B-L, and $k-\varepsilon$ compressibility turbulence models

The observed differences among the models are confirmed by Figure 4, where the boundary layer separation distance from the jet is shown as a function of Mach number. In the case of taking into account compressibility effect (Figure 4, line 3) the curve of the separation distance is located between $k-\varepsilon$ and Baldwin-Lomax model lines. The curves show that the sizes of counter-rotating primary and secondary upstream vortices in the region upstream of the jet differ.

Fig. 5. (a-k) shows contours of the constant mass concentrations of hydrogen for different M_∞. The penetration and mixing of hydrogen is performed in accordance with the physics of flows, consequently the depth of penetration hydrogen grows with increased M_∞.

Contour plots of turbulent characteristics of the flow are shown in Figures 6 and 7. Figure 6 shows comparison of turbulence kinetic energy k with turbulence energy dissipation ε calculated by $k-\varepsilon$ model for various freestream Mach numbers, while Figure 7 shows the same comparison computed with the $k-\varepsilon$ compressibility model.

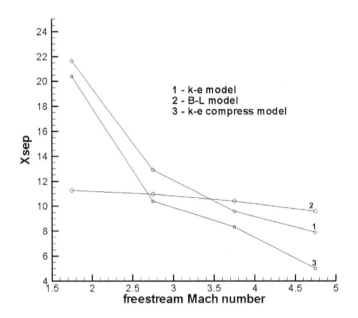

Fig. 4. Separation length as a function of freestream Mach number.

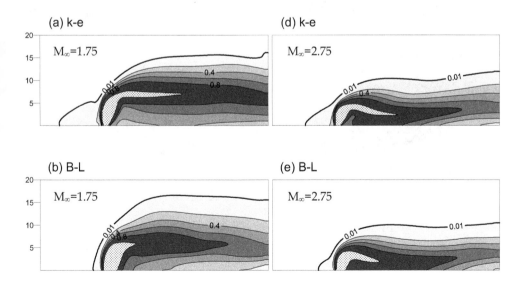

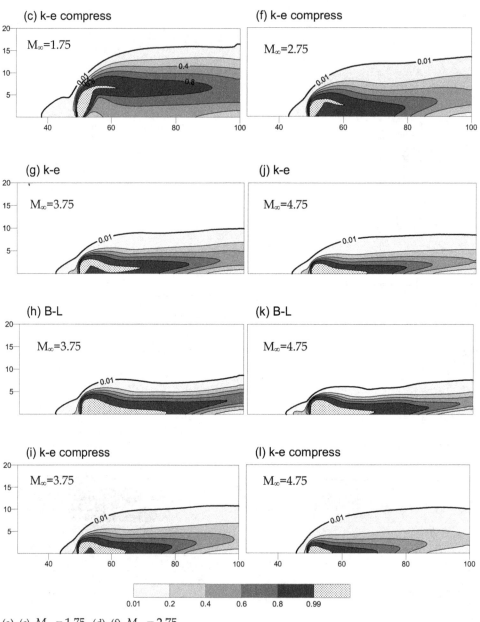

(a)–(c) $M_\infty = 1.75$, (d)–(f) $M_\infty = 2.75$

(g)–(i) $M_\infty = 3.75$, (j)–(l) $M_\infty = 4.75$

Fig. 5. Mass fraction contour of hydrogen.

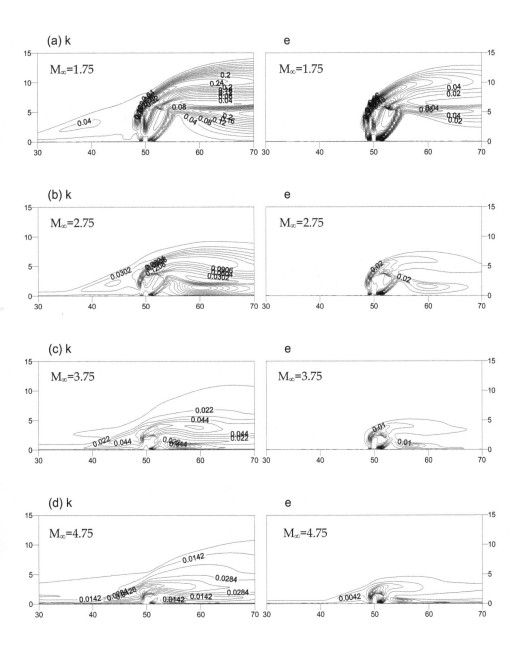

Fig. 6. Comparison of $k - \varepsilon$ contours for various freestream Mach numbers, calculated by the $k - \varepsilon$ model.

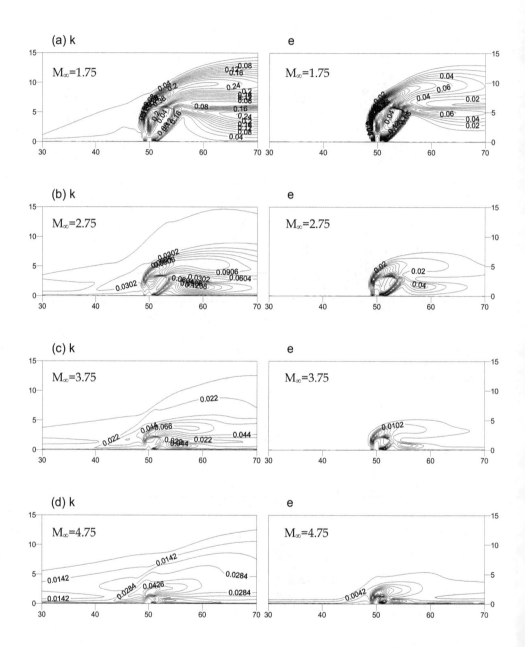

Fig. 7. Comparison of $k-\varepsilon$ contours for various freestream Mach numbers, calculated by the $k-\varepsilon$ compressibility model.

The dynamics of reduction of dimensionless residual norm of gas density for the implicit method is shown in the Figure 8. The figure shows that increasing the subsonic zones (smaller M_∞) degrades the convergence. Apparently, numerical simulation with smaller of M_∞ will need optimization in order to provide better convergence.

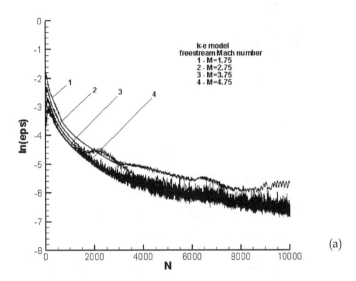

(a)

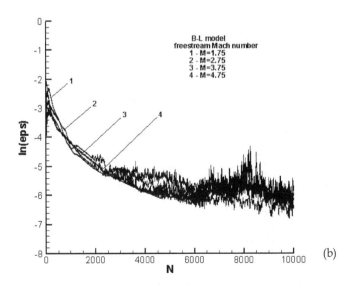

(b)

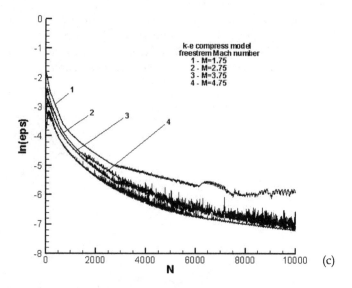

Fig. 8. Dimensionless residual norm for the gas density vs. the number of iterations (a)-
$k-\varepsilon$ model, (b)–B-L model, (c)- $k-\varepsilon$ compressibility model.

The influence of the pressure ratio on the upstream flow structure was also studied. An analysis of the influence of the jet pressure ratio on the bow shock wave shows that the angle of inclination of the latter increases with increasing pressure ratio behind the shock wave and ahead of it. The effect of the jet pressure ratio on the separation-region length is illustrated in Figure 9 for $M_\infty = 3.75$. Here results are represented which were calculated by the convergence of the dimensionless residual norm for gas density $\ln \varepsilon = -5$.

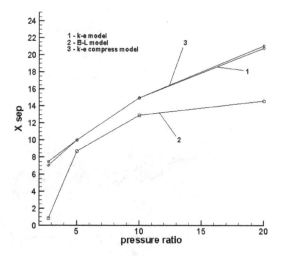

Fig. 9. Separation length as a function of static pressure ratio.

7. Conclusion

In the present work, the numerical model and computer code have been developed for a $k-\varepsilon$ turbulent model that includes compressibility correction to simulate turbulent supersonic air flow with transfer injection of the hydrogen from slots of the walls. The numerical method is based on the ENO scheme.

Comparison is made between solution by using algebraic Baldwin-Lomax's turbulence model and $k-\varepsilon$ model. The influence of the compressibility effect on the shock wave structure is studied.

The interaction of the supersonic air flow with transfer injection of jets depending on the freestream Mach number and pressure ratio have been studied. In spite of similarity in the shock wave structure for both turbulence model, the numerical experiments have shown that the sizes of counter-rotating primary and secondary upstream vortices in the region upstream of the jet differ.

The developed computer code allows to study of mixing flowfield behavior with various injector geometries and reacting flows which is important problems in the design of supersonic combustion ramjet (scramjet) engines.

8. References

[1] Drummond J. P. and Weidner E. H., "Numerical Study of a Scramjet Engine Flowfield", AIAA Journal, Vol. 20, No.9, 1982, pp. 1182-1187.

[2] Weidner E. H. and Drummond J. P., "Numerical Study of Staged Fuel Injection for Supersonic Combustion", AIAA Journal, Vol. 20, No.10, 1982, pp. 1426-1431.

[3] Clarence F. Chenault and Philip S. Beran, "k-e and Reynolds Stress Turbulence Model Comparisons for Two-Dimensional Injection Flows", AIAA Journal, Vol. 36, No. 8, 1998, pp. 1401-1412.

[4] Grasso F. and Magi V., "Simulation of Transverse Gas Injection in Turbulent Supersonic Air Flows", AIAA Journal, Vol. 33, No. 1, 1995, pp. 56-62.

[5] Mohammad Ali, Toshi Fujiwara, Joseph E. Leblanc, "Influence of main flow inlet configuration on mixing and flameholding in transverse injection into supersonic airstream", International Journal of Engineering Science, 2000, pp. 1161-1180.

[6] Mohammad Ali, S. Ahmed, A. K. M. Sadrul Islam, "The Two-Dimensional Supersonic Flow and Mixing with a Perpendicular Injection in a Scramjet Combustor", Journal of Thermal Science, Vol. 12, No.4, 2003, pp. 372-380.

[7] Shuen J. S. and Yoon S., "Numerical Study of Chemically Reacting Flows Using a Lower-Upper Symmetric Successive Overrelaxation Scheme", AIAA Journal, Vol. 27, No. 12, 1989, pp. 1752-1760.

[8] P. Bruel and A. Zh. Naimanova, "Computation of the normal injection of a hydrogen jet into a supersonic air flow", Thermophysics and Aeromechanics, Vol.17, No. 4, 2010, pp. 531-542.

[9] Beketaeva A.O., Naimanova A.Zh. Numerical simulation of a supersonic flow with transverse injection of jets // Journal of Applied Mechanics and Technical physics, 2004. Vol.45, №3. P.367-374.

[10] D. P. Rizzetta (1995) Numerical Investigation of Supersonic Wing-Tip Vortices. 26th AIAA Fluid Dynamics Conference.

[11] Rizzetta D.P., "Numerical Simulation of Slot Injection into a Turbulent Supersonic Stream", AIAA Journal, Vol.30, No.10, 1992, pp. 2434-2439.

[12] Kee R.J., Rupley F.M., Miller J.A. CHEMKIN-II: a Fortran chemical kinetic package for the analysis of gas-phase chemical kinetics // SANDIA Report SAND89-8009. – 1989.

[13] J. O. Hirschfelder, Ch. F. Curtiss, R. B. Bird, "Molecular Theory of Gases and Liquids", Moscow, 1961.

[14] Dean R. Eklund and J. Philip Drummond, NASA Langley Research Center, Hampton, Virginia and H.A. Hassan, North Carolina State University, Raleigh, North Carolina. Calculation of Supersonic Turbulent Reacting Coaxial Jets // AIAA Journal. Vol.28, NO.9, September 1990.

[15] SUN De-chuan, HU Chun-bo, CAI Ti-min. Computation of Supersonic Turbulent Flowfield with Transverse Injection // Applied Mathematics and Mechanics. English Edition. Vol.23, No 1, Jan 2002. P. 107-113.

[16] Kee R.J., Rupley F.M., Miller J.A. CHEMKIN-II: a Fortran chemical kinetic package for the analysis of gas-phase chemical kinetics // SANDIA Report SAND89-8009. – 1989.

[17] Harten A., Engquist B., Osher S., Chakravarthy S.R. Uniformly High Order Accurate Non-oscillatory Schemes, III // Journal of Computational Physics. – 1987. – Vol.71. – P.231-303.

[18] Harten A., Osher S., Engquist B., Chakravarthy S.R. Some Results on Uniformly High-Order Accurate Essentially Non-oscillatory Schemes // Applied Num. Math. – 1986. – Vol.2. – P.347-377.

Part 3

Advances in Classical Approaches

3

Parallel Accelerated Group Iterative Algorithms in the Solution of Two-Space Dimensional Diffusion Equations

Norhashidah Hj Mohd Ali and Foo Kai Pin
Universiti Sains Malaysia
Malaysia

1. Introduction

Diffusion equations are mathematical models which explain how the concentration of one or more substances distributed in space is altered by a diffusion process which causes the substances to spread out over a surface in space. For a normal diffusion process, the flux of particles into one region must be the sum of particle flux flowing out of the surrounding regions. From Fick's first law, this can be represented mathematically by the following diffusion equation

$$\frac{\partial u}{\partial t} = \nabla.(D\nabla u) . \qquad (1)$$

If the diffusion coefficient is constant in space, then

$$\frac{\partial u}{\partial t} = D\nabla^2 u \qquad (2)$$

here ∇^2 is the normal Laplacian. The values u and D take on different meanings in different situations: in particle diffusion, u is interpreted as a concentration and D as a diffusion coefficient; while in heat diffusion, u is the temperature and D is the thermal conductivity.

The application of the finite difference methods for solving time-dependent Partial Differential Equations(PDEs) such as this, at any particular time level, yields a system of linear simultaneous equations of the form

$$A\underline{u} = \underline{b} \qquad (3)$$

where iterative methods are normally more feasible in solving the system due to the sparsity of the matrix A. The applicability of explicit group methods, in which several unknowns are connected together in the iteration formula resulting in a sub-system that must be solved before any one of the them can be determined, have been investigated on solving these types of PDEs. In their early work, Evans and Abdullah (1983b) generated single-step, one-parameter families of finite difference approximations to the heat equation in one space dimension by coupling in groups of two the values of the approximations obtained by known asymmetric formulas at adjacent grid points at the advanced time level. The resulting equations are implicit but they can be easily converted to explicit form. The method was shown to possess unconditional stability with good accuracies. Evans and Abdullah (1983a) also developed the group explicit method for the solution the two dimensional diffusion where a general two-level six point finite difference approximation was developed to solve the parabolic equation. The method was further developed as the Alternating Group Explicit (AGE) method (Evans & Sahimi, 1988), which is an analogue to the famous Alternating Direction Implicit(ADI) method but has the advantage of being explicit and thus very easy to parallelise. The emergence of newer explicit group methods on skewed or rotated grids with promising and improved results was greatly observed since the early 1990s. Among them are the works of Abdullah (1991) who developed the four-point Explicit Decoupled Group (EDG) by discretising the PDEs on skewed grids. This method was shown to require less computational time with the same order of accuracies than the Explicit Group (EG) method pioneered by Yousif and Evans (1986) in solving the Poisson model problem. A few years on, Othman and Abdullah (2000) modified the formulation of the EG method by deriving formulas based on the centred five points approximation formula with the grid spacing h and $2h$, and the rotated five points approximation formula to come up with the improved modified four-point EG which was shown to exhibit lesser computational effort than the existing EG and EDG. Since then, active research has been conducted to investigate the capabilities of the variants of these group relaxation methods in improving the standard or traditional algorithms in solving several types of PDEs. This includes the work of Ali and Lee (2007) who derived the Accelerated OverRelaxation (AOR) variant of the EDG group scheme in the solution of elliptic equation where its performance results were compared with the EG (AOR) proposed by Martins et al. (2002). The new EDG (AOR) scheme requires less execution time than the existing EG (AOR) method where the gain in speed of EDG (AOR) method over the EG (AOR) method ranges from approximately 51% to 59%. The performance analysis of the parallel algorithms of these EG and EDG schemes were also established in Ng and Ali (2008) where the algorithms turn out to be efficient solvers for the steady-state elliptic equation on distributed memory multicomputer with high scalability.

In this chapter we shall present the formulation of new explicit group algorithms intended for solving the two dimensional time-dependent diffusion equation. A novel approach of using four points group strategy, implemented on different spacing stencils incorporated with the AOR technique, is used in the formulation. Explainations on how the methods need to be reconfigurated mathematically as to be successfully ported to run on a message-passing parallel computer system is also presented.

2. Formulations of group methods

We consider the finite difference discretization schemes for solving the two dimensional diffusion equation of the form

$$\frac{\partial u}{\partial t} = \nabla^2 u + f(x,y,t) \tag{4}$$

with a specified initial and boundary conditions on a unit square with spacings $\Delta x = \Delta y = h = 1/n$ in both directions x and y, with $x_i = x_0 + ih$, $y_j = y_0 + jh$ $(i,j = 0,1,2,...,n)$, $t = k\Delta t$ $(k = 0, 1, 2, ...)$; here, f is a real continuous function. One commonly used implicit finite difference scheme based on the centred difference in time and space formulation about the point $(i,j,k+1/2)$ is the *Crank-Nicolson* scheme which transform (4) into

$$(1+2r)u_{i,j,k+1} - \frac{r}{2}u_{i-1,j,k+1} - \frac{r}{2}u_{i+1,j,k+1} - \frac{r}{2}u_{i,j-1,k+1} - \frac{r}{2}u_{i,j+1,k+1} =$$

$$(1-2r)u_{i,j,k} + \frac{r}{2}u_{i-1,j,k} + \frac{r}{2}u_{i+1,j,k} + \frac{r}{2}u_{i,j-1,k} + \frac{r}{2}u_{i,j+1,k} + \Delta t f_{i,j,k+\frac{1}{2}}$$

(5)

where $r = \Delta t / h^2$. Based on this approximation, several group schemes have been constructed (Ali, 1998). The Explicit Group (EG) method, for example, was formulated by taking the iteration process in groups of four points. At each time level $(k+1)$, the mesh points are grouped in blocks of four points (i,j), $(i+1,j)$, $(i+1,j+1)$ and $(i,j+1)$ and equation (5) is applied to each of these points resulting in the following (4x4) system:

$$\begin{bmatrix} 1+2r & -\frac{r}{2} & 0 & -\frac{r}{2} \\ -\frac{r}{2} & 1+2r & -\frac{r}{2} & 0 \\ 0 & -\frac{r}{2} & 1+2r & -\frac{r}{2} \\ -\frac{r}{2} & 0 & -\frac{r}{2} & 1+2r \end{bmatrix} \begin{bmatrix} u_{i,j,k+1} \\ u_{i+1,j,k+1} \\ u_{i+1,j+1,k+1} \\ u_{i,j+1,k+1} \end{bmatrix} = \begin{bmatrix} rhs_{i,j} \\ rhs_{i+1,j} \\ rhs_{i+1,j+1} \\ rhs_{i,j+1} \end{bmatrix}$$

(6)

which may be solved explicitly in groups of four points as

$$\begin{bmatrix} u_{i,j,k+1} \\ u_{i+1,j,k+1} \\ u_{i+1,j+1,k+1} \\ u_{i,j+1,k+1} \end{bmatrix} =$$

$$\frac{1}{2(1+r)(1+2r)+(1+3r)} \begin{bmatrix} 7r^2+8r+2 & r(1+2r) & r^2 & r(1+2r) \\ r(1+2r) & 7r^2+8r+2 & r(1+2r) & r^2 \\ r^2 & r(1+2r) & 7r^2+8r+2 & r(1+2r) \\ r(1+2r) & r^2 & r(1+2r) & 7r^2+8r+2 \end{bmatrix} \begin{bmatrix} rhs_{i,j} \\ rhs_{i+1,j} \\ rhs_{i+1,j+1} \\ rhs_{i,j+1} \end{bmatrix}, (7)$$

where

$$rhs_{i,j} = \frac{r}{2}\left(u_{i-1,j,k+1} + u_{i,j-1,k+1}\right) + \frac{r}{2}\left(u_{i-1,j,k} + u_{i,j-1,k} + u_{i+1,j,k} + u_{i,j+1,k}\right)$$
$$+ (1-2r)u_{i,j,k} + \Delta t f_{i,j,k+\frac{1}{2}}$$

$$rhs_{i+1,j} = \frac{r}{2}\left(u_{i+2,j,k+1} + u_{i+1,j-1,k+1}\right) + \frac{r}{2}\left(u_{i,j,k} + u_{i+2,j,k} + u_{i+1,j-1,k} + u_{i+1,j+1,k}\right)$$
$$+ (1-2r)u_{i+1,j,k} + \Delta t f_{i+1,j,k+\frac{1}{2}}$$

$$rhs_{i+1,j+1} = \frac{r}{2}\left(u_{i+2,j+1,k+1} + u_{i+1,j+2,k+1}\right) + \frac{r}{2}\left(u_{i,j+1,k} + u_{i+2,j+1,k} + u_{i+1,j,k} + u_{i+1,j+2,k}\right)$$
$$+ (1-2r)u_{i+1,j+1,k} + \Delta t f_{i+1,j+1,k+\frac{1}{2}}$$

$$rhs_{i,j+1} = \frac{r}{2}\left(u_{i-1,j+1,k+1} + u_{i,j+2,k+1}\right) + \frac{r}{2}\left(u_{i-1,j+1,k} + u_{i,j,k} + u_{i+1,j+1,k} + u_{i,j+2,k}\right)$$
$$+ (1-2r)u_{i,j+1,k} + \Delta t f_{i,j+1,k+\frac{1}{2}}$$

The method proceed with iterative evaluation of solutions in blocks of four points respectively using these formulas throughout the whole net region until convergence is achieved.

Using another type of discretization, which we called the *rotated* finite difference approximation:

$$
\begin{aligned}
\frac{u_{i,j,k+1} - u_{i,j,k}}{\Delta t} &= \frac{1}{2}\left[\frac{u_{i-1,j-1,k+1} - 2u_{i,j,k+1} + u_{i+1,j+1,k+1}}{2\Delta x^2} + \frac{u_{i-1,j-1,k} - 2u_{i,j,k} + u_{i+1,j+1,k}}{2\Delta x^2} \right] \\
&+ \frac{1}{2}\left[\frac{u_{i-1,j+1,k+1} - 2u_{i,j,k+1} + u_{i+1,j-1,k+1}}{2\Delta y^2} + \frac{u_{i-1,j+1,k} - 2u_{i,j,k} + u_{i+1,j-1,k}}{2\Delta y^2} \right] + f_{i,j,k+\frac{1}{2}}
\end{aligned}
\tag{8}
$$

another group scheme was formulated called the Explicit Decoupled Group (EDG) method. This approximation is obtained by using the Taylor series expansion of the solution u at appropriate grid points where the resulting computational stencil is 45^0 clockwise rotated from the stencil of the standard *Crank-Nicolson* (5). At each time level ($k+1$), the mesh points are grouped in blocks of four points, (i,j), $(i+1,j+1)$, $(i+1,j)$ and $(i,j+1)$, and the rotated finite difference approximation (8) is applied to the u values at each of these points resulting in a (4x4) system which may be decoupled into two 2x2 matrices of the following form (Ali, 1998):

$$
\cdot \begin{bmatrix} u_{i,j,k+1} \\ u_{i+1,j+1,k+1} \end{bmatrix} = \frac{16}{15r^2 + 32r + 16} \begin{bmatrix} 1+r & \dfrac{r}{4} \\ \dfrac{r}{4} & 1+r \end{bmatrix} \begin{bmatrix} rhs_{i,j} \\ rhs_{i+1,j+1} \end{bmatrix},
\tag{9}
$$

and

$$
\begin{bmatrix} u_{i+1,j,k+1} \\ u_{i,j+1,k+1} \end{bmatrix} = \frac{16}{15r^2 + 32r + 16} \begin{bmatrix} 1+r & \dfrac{r}{4} \\ \dfrac{r}{4} & 1+r \end{bmatrix} \begin{bmatrix} rhs_{i+1,j} \\ rhs_{i,j+1} \end{bmatrix},
\tag{10}
$$

where

$$
\begin{aligned}
rhs_{i,j} &= \frac{r}{4}\left(u_{i-1,j-1,k+1} + u_{i+1,j-1,k+1} + u_{i-1,j+1,k+1} \right) + \frac{r}{4}\left(u_{i-1,j-1,k} + u_{i+1,j-1,k} + u_{i-1,j+1,k} + u_{i+1,j+1,k} \right) \\
&+ (1-r)u_{i,j,k} + \Delta t f_{i,j,k+\frac{1}{2}}
\end{aligned}
$$

$$
\begin{aligned}
rhs_{i+1,j+1} &= \frac{r}{4}\left(u_{i+2,j+2,k+1} + u_{i+2,j,k+1} + u_{i,j+2,k+1} \right) + \frac{r}{4}\left(u_{i+2,j+2,k} + u_{i+2,j,k} + u_{i,j+2,k} + u_{i,j+2,k} \right) \\
&+ (1-r)u_{i+1,j+1,k} + \Delta t f_{i+1,j+1,k+\frac{1}{2}}
\end{aligned}
$$

$$
\begin{aligned}
rhs_{i+1,j} &= \frac{r}{4}\left(u_{i,j-1,k+1} + u_{i+2,j-1,k+1} + u_{i+2,j+1,k+1} \right) + \frac{r}{4}\left(u_{i,j-1,k} + u_{i+2,j-1,k} + u_{i,j+1,k} + u_{i+2,j+1,k} \right) \\
&+ (1-r)u_{i+1,j,k} + \Delta t f_{i+1,j,k+\frac{1}{2}}
\end{aligned}
$$

$$
\begin{aligned}
rhs_{i,j+1} &= \frac{r}{4}\left(u_{i-1,j,k+1} + u_{i+1,j+2,k+1} + u_{i-1,j+2,k+1} \right) + \frac{r}{4}\left(u_{i-1,j,k} + u_{i+1,j+2,k} + u_{i-1,j+2,k} + u_{i+1,j,k} \right) \\
&+ (1-r)u_{i,j+1,k} + \Delta t f_{i,j+1,k+\frac{1}{2}}.
\end{aligned}
\tag{11}
$$

The EDG method requires less computing time whilst maintaining the same order of accuracies as the original EG method (Ali, 1998). Based on approximation formula (8) derived on different grid spacings, new modified group explicit methods will be formulated in combination of the Accelerated Over-Relaxation (AOR) technique. The next sub-sections will elaborate on the formulations of these methods.

2.1 Modified explicit group (MEG) AOR method

We consider the standard five-point formula for the diffusion equation on Ω_{2h} grid:

$$u_{i,j}^{(m+1)} = \frac{1}{1+2r}\left(\frac{r}{2}(u_{i-2,j}^{(m)} + u_{i+2,j}^{(m)} + u_{i,j-2}^{(m)} + u_{i,j+2}^{(m)} + v_{i-2,j} + v_{i+2,j} + v_{i,j-2} + v_{i,j+2}) + (1-2r)v_{i,j} + \Delta tF_{i,j}\right), \quad (12)$$

$$r = \frac{\Delta t}{4h^2}.$$

Here, u_{ij} is the value of u at the current time level $(k+1)$, v_{ij} is the value of u from the previous time level (k), while m is the iteration level. F_{ij} is the value of f at the point (i,j) at time level $(k+1)$. We begin by dividing the grid points in the solution domain into 3 types of points, indicated by $\square$, $\triangle$, $\bullet$, and arranged in a specific alternate ordering, as shown in Fig. 1. For the iterations, we consider the points indicated by $\bullet$ in Fig. 2. Similar to the EG method described in the previous section, we may apply equation (12) to groups of four points of the iterative points to produce a 4x4 MEG formula. The convergence of this method may be improved by the introduction of the AOR technique (Hadjidimos, 1978) where drastic improvement in convergence can be obtained by choosing suitable relaxation parameters in its formula. The idea in AOR technique is to apply an extrapolation of a two-parameter Successive OverRelaxation (SOR)-type iterative procedure in the formula.

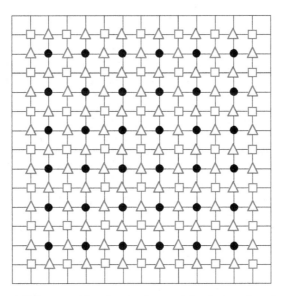

Fig. 1. Construction of different grid points in the spatial solution domain at time level $k+1$.

The two parameters may be exploited which result in methods which will converge faster than any other method of the same type. Using this idea, we can introduce the over-relaxation parameters ω and R into the 4x4 MEG formula as a way to further accelerate the convergence of the iterative method scheme as the following:

$$u_{i,j}^{(m+1)} = s_3[w(s_4 b_5 + s_5 b_6 + s_6 b_7 + s_5 b_8) + R(s_4 t_1 + s_5\{t_2 + t_3\})] + (1-w)u_{i,j}^{(m)}$$

$$u_{i+2,j}^{(m+1)} = s_3[w(s_5 b_5 + s_4 b_6 + s_5 b_7 + s_6 b_8) + R(s_5 t_1 + s_4 t_2 + s_6 t_3)] + (1-w)u_{i+2,j}^{(m)}$$

$$u_{i+2,j+2}^{(m+1)} = s_3[w(s_6 b_5 + s_5 b_6 + s_4 b_7 + s_5 b_8) + R(s_6 t_1 + s_5\{t_2 + t_3\})] + (1-w)u_{i+2,j+2}^{(m)}$$

$$u_{i,j+2}^{(m+1)} = s_3[w(s_5 b_5 + s_6 b_6 + s_5 b_7 + s_4 b_8) + R(s_5 t_1 + s_6 t_2 + s_4 t_3)] + (1-w)u_{i,j+2}^{(m)}$$

(13)

where

$$b_1 = s_2 v_{i,j} + \Delta t F_{i,j}$$

$$b_2 = s_2 v_{i+2,j} + \Delta t F_{i+2,j}$$

$$b_3 = s_2 v_{i+2,j+2} + \Delta t F_{i+2,j+2}$$

$$b_4 = s_2 v_{i,j+2} + \Delta t F_{i,j+2}$$

$$b_5 = s_1(u_{i-2,j}^{(m)} + u_{i,j-2}^{(m)} + v_{i-2,j} + v_{i+2,j} + v_{i,j-2} + v_{i,j+2}) + b_1$$

$$b_6 = s_1(u_{i+4,j}^{(m)} + u_{i+2,j-2}^{(m)} + v_{i,j} + v_{i+4,j} + v_{i+2,j-2} + v_{i+2,j+2}) + b_2$$

$$b_7 = s_1(u_{i+4,j+2}^{(m)} + u_{i+2,j+4}^{(m)} + v_{i,j+2} + v_{i+4,j+2} + v_{i+2,j} + v_{i+2,j+4}) + b_3$$

$$b_8 = s_1(u_{i-2,j+2}^{(m)} + u_{i,j+4}^{(m)} + v_{i-2,j+2} + v_{i+2,j+2} + v_{i,j} + v_{i,j+4}) + b_4$$

$$s_1 = 0.5r$$

$$s_2 = 1 - 2r$$

$$s_3 = \frac{1}{2(1+r)(1+2r)(1+3r)}$$

$$s_4 = 7r^2 + 8r + 2$$

$$s_5 = r(1+2r)$$

$$s_6 = r^2$$

$$t_1 = s_1(u_{i-2,j}^{(m+1)} - u_{i-2,j}^{(m)} + u_{i,j-2}^{(m+1)} - u_{i,j-2}^{(m)})$$

$$t_2 = s_1(u_{i+2,j-2}^{(m+1)} - u_{i+2,j-2}^{(m)})$$

$$t_3 = s_1(u_{i-2,j+2}^{(m+1)} - u_{i-2,j+2}^{(m)}).$$

(14)

Unlike SOR, there is no general formula to determine the parameters R and w. But according to Hadjidimos (1978), the parameter R is normally chosen to be close to the value ω obtained from the corresponding SOR technique which give the least number of iterations. We can then define the four points MEG (AOR) method for diffusion equation as the following:

Algorithm 1

1. Divide the grid points into points of type ●, △ and □ at level $k+1$ as shown in Fig. 1.
2. Set the initial guess for the iterations.
3. Use Equations (13)-(14) to evaluate the solution of points of type ● iteratively at level $k+1$.
4. Check the convergence. If the iterations converge, go to step 5. Otherwise, repeat step 3 until convergence is achieved.
5. After the solution at points of type ● converge, the converged values are then adopted as the initial guess for the next time level.
6. Then, repeat steps 1 to 5 until the solutions at all the required time levels have been obtained.
7. For the solutions at the remaining points at level $k+1$ (Fig. 1), compute them directly once according to the following sequence:

a) For points of type □, use the *rotated* five points approximation formula on the $\Omega_{\sqrt{2}h}$ grid:

$$u_{i,j} = \frac{1}{1+2r_2}(\frac{r_2}{2}(u_{i-1,j-1} + u_{i+1,j-1} + u_{i-1,j+1} + u_{i+1,j+1}$$

$$+v_{i-1,j-1} + v_{i+1,j-1} + v_{i+1,j-1} + v_{i+1,j+1}) + (1-2r_2)v_{i,j} + \Delta t F_{i,j}) , r_2 = \frac{\Delta t}{2h^2} \qquad (15)$$

b)For points of type △, use the standard five points approximation formula on the Ω_h grid:

$$u_{i,j} = \frac{1}{1+2r_3} X$$

$$\left(\frac{r_3}{2}(u_{i-1,j} + u_{i+1,j} + u_{i,j-1} + u_{i,j+1} + v_{i-1,j} + v_{i+1,j} + v_{i,j-1} + v_{i,j+1}) + (1-2r_3)v_{i,j} + \Delta t F_{i,j}\right) \qquad (16)$$

Here,

$$r_3 = \frac{\Delta t}{h^2} .$$

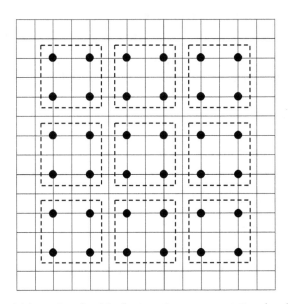

Fig. 2. Grid points which are involved in the iterative process at time level $k+1$.

2.2 Modified Explicit Decoupled Group (MEDG) AOR method

To formulate the MEDG (AOR) method for the diffusion equation, we consider the *rotated* five-point approximation formula for the diffusion equation with $2h$ spacing:

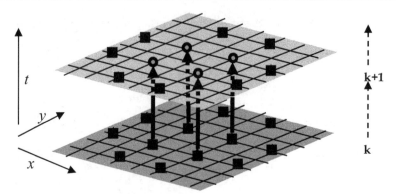

Fig. 3. Points involved in the updates of solutions in 4 points MEG (AOR).

$$u_{i,j}^{(m+1)} = \frac{1}{1+2r}\left(\frac{r}{2}(u_{i-2,j-2}^{(m)} + u_{i+2,j-2}^{(m)} + u_{i-2,j+2}^{(m)} + u_{i+2,j+2}^{(m)}\right.$$

$$\left. + v_{i-2,j-2} + v_{i+2,j-2} + v_{i-2,j+2} + v_{i+2,j+2}) + (1-2r)v_{i,j} + \Delta t F_{i,j}\right) \quad (17)$$

Here,

$$r = \frac{\Delta t}{4h^2}.$$

The method is constructed by firstly dividing the grid points into 4 types of points in a specific alternate ordering as shown in Fig. 5 in a unit square domain with $n=14$. The MEDG (AOR) formula for diffusion equation can then be obtained by applying equation (17) to groups of points of type ● in the solution domain. This application will produce a 4x4 system which can be inverted and rewritten in explicit decoupled form of two equations:

$$u_{i,j}^{(m+1)} = s_1[w(s_3 b_3 + s_2 b_4) + R(s_3 t_1)] + (1-w)u_{i,j}^{(m)}$$

$$u_{i+2,j+2}^{(m+1)} = s_1[w(s_2 b_3 + s_3 b_4) + R(s_2 t_1)] + (1-w)u_{i+2,j+2}^{(m)} \quad (18)$$

where

$$b_1 = s_4 v_{i,j} + \Delta t F_{i,j}$$

$$b_2 = s_4 v_{i+2,j} + \Delta t F_{i+2,j+2}$$

$$b_3 = s_2(u_{i-2,j-2}^{(m)} + u_{i+2,j-2}^{(m)} + u_{i-2,j+2}^{(m)} + v_{i-2,j-2} + v_{i+2,j-2} + v_{i-2,j+2} + v_{i+2,j+2}) + b_1$$

$$b_4 = s_2(u_{i+4,j+4}^{(m)} + u_{i+4,j}^{(m)} + u_{i,j+4}^{(m)} + v_{i+4,j+4} + v_{i+4,j} + v_{i,j+4} + v_{i,j}) + b_2$$

$$t_1 = u_{i-2,j-2}^{(m+1)} - u_{i-2,j-2}^{(m)} + u_{i+2,j-2}^{(m+1)} - u_{i+2,j-2}^{(m)} + u_{i-2,j+2}^{(m+1)} - u_{i-2,j+2}^{(m)}$$

$$s_1 = \frac{16}{15r^2 + 32r + 16}$$

$$s_2 = 0.25r \quad (19)$$

$$s_3 = 1 + r$$

$$s_4 = 1 - r$$

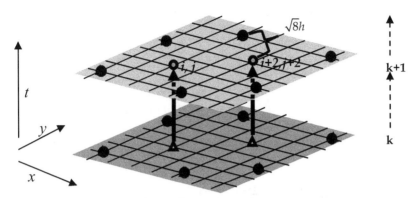

Fig. 4. Group of points involved in the iterative process with the spacing of 2*h* for MEDG (AOR) at time level *k*+1 and *k*.

The resulting grid or the computational molecule to update at the two points can be viewed in Fig. 4 with a mesh size 2*h*. From Fig. 5, it is obvious that the evaluation of equations (18) - (19) involves only points of type ●. This means that by using the approximation formulas (18)-(19), it is easy to see that the black filled points are linked only to the same type of points. Thus the iterative procedure involving these formulas can be performed independently of the other type of points. We can then formulate the four points MEDG (AOR) method as in **Algorithm 2**:

Algorithm 2

1. Divide the grid points at layer *k*+1 into points of type ●, O ,△ and □ as shown in Fig. 5.
2. Set the initial guess for the iterations.
3. Evaluate the solution at the points of type ● iteratively at layer *k*+1 by using equations (18)-(19).
4. Check the convergence. If the iterations converge, go to step 5. Otherwise, repeat step 2 and step 3 until convergence is achieved.
5. After the solutions at points of type ● converge, the converged points are then adopted as initial guess for the next time level.
6. Then, repeat steps 1 to 5 until the solutions at all the required time levels have been obtained.
7. For the solutions at the remaining points at layer *k*+1 (Fig. 5), compute them directly once according to the following sequence:
   a. For points of type O, use the standard five points approximation formula on the Ω_{2h} grid:

$$u_{i,j} = \frac{1}{1+2r_1}(\frac{r_1}{2}(u_{i-2,j} + u_{i+2,j} + u_{i,j-2} + u_{i,j+2}$$

$$+v_{i-2,j} + v_{i+2,j} + v_{i,j-2} + v_{i,j+2}) + (1 - 2r_1)v_{i,j} + \Delta t F_{i,j}) \quad r_1 = \frac{\Delta t}{4h^2} \tag{20}$$

b. For points of type □, use the *rotated* five points approximation formula on the $\Omega_{\sqrt{2}h}$ grid:

$$u_{i,j} = \frac{1}{1+2r_2}(\frac{r_2}{2}(u_{i-1,j-1}+u_{i+1,j-1}+u_{i-1,j+1}+u_{i+1,j+1}$$

$$+v_{i-1,j-1}+v_{i+1,j-1}+v_{i+1,j-1}+v_{i+1,j+1})+(1-2r_2)v_{i,j}+\Delta tF_{i,j}) , \; r_2 = \frac{\Delta t}{2h^2} \qquad (21)$$

c. For points of type △, use the standard five points approximation formula on the Ω_h grid:

$$u_{i,j} = \frac{1}{1+2r_3}(\frac{r_3}{2}(u_{i-1,j}+u_{i+1,j}+u_{i,j-1}+u_{i,j+1}$$

$$+v_{i-1,j}+v_{i+1,j}+v_{i,j-1}+v_{i,j+1})+(1-2r_3)v_{i,j}+\Delta tF_{i,j}) , \; r_3 = \frac{\Delta t}{h^2} \qquad (22)$$

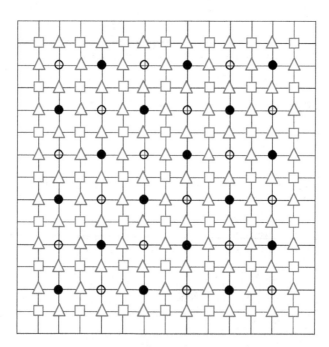

Fig. 5. The discretized domain of the four points MEDG (AOR) method at time level $k+1$.

3. Numerical experiments of the sequential group methods

In order to verify the performance of the proposed methods which were shown in previous sections, the algorithms were tested on the following model problem:

$$\frac{\partial u}{\partial t} = \frac{\partial^2 u}{\partial x^2} + \frac{\partial^2 u}{\partial y^2} + \sin x \sin y e^{-t} - 4 \tag{23}$$

with Dirichlet boundary conditions satisfying its exact solution

$$u(x,y,t) = \sin x \sin y e^{-t} + x^2 + y^2,$$

$(x,y) \in \partial\Omega$, $\partial\Omega$ is the boundary of the unit square Ω. The model equation (23) is of the form

$$\frac{\partial u}{\partial t} = \frac{\partial^2 u}{\partial x^2} + \frac{\partial^2 u}{\partial y^2} + g(x,y,t,u)$$

and is normally called a reaction-diffusion equation which models the movement of basic particles in sciences (for example, heat transfer, growth population, or dilution of chemical in water) in a region Ω. Here,

$$\frac{\partial^2 u}{\partial x^2} + \frac{\partial^2 u}{\partial y^2}$$

is the diffusion term which describes the movement of the particles and the remaining term at the right hand side of equation (23) (i.e. $g(x,y,t,u)$) is the reaction term which describes changes (due to birth, death, chemical reactions, etc.) occuring inside the region (or habitat). For this numerical experiment, we purposely find a model problem which has an exact solution to ensure that the proposed methods yield correct results. To terminate the iteration process, the relative error test, i.e. $Error = abs(u_{ij}^{(m+1)} - u_{ij}^{(m)}) / (1 + abs(u_{ij}^{(m+1)}))$, was used as the convergence test with tolerance $\varepsilon = 1.0 \times 10^{-6}$. As described in Section 2.1, the over-relaxation parameters, R and w, need to be found which give the best convergence rates for the proposed schemes. To achieve this, we obtained the values of w for the corresponding SOR scheme, and then the value of R was found by running the experiments using these specific values of w which gave the least number of iterations. Different grid sizes of $n = 82, 102, 122,$ 142, 162, 182 and 202 were chosen to record the total iteration counts (*Iter*) at all time levels and computer timings (t) of the group AOR methods. The value of $\Delta t = 0.0005$ with 1000 time levels was used to run the programs.

The numerical results of the proposed MEG(AOR) and MEDG (AOR) methods together with the original explicit group methods EG(AOR), EDG(AOR) are tabulated in Tables 1 to 2. The point AOR method which uses the existing traditional *Crank-Nicolson* scheme (5) accelerated with the AOR technique is also shown in Table 3 for comparison purposes . The value of R was chosen experimentally to be close to the value of w as depicted in the tables. All of the methods tested are of second order accuracies so that the results they produce are of similar accuracies as seen in Table 4. From Tables 1 and 2, it can be seen that between EG(AOR) and EDG(AOR), the latter has better rates of convergence which is consistent with the results in Ali and Lee (2007) for the elliptic problem. The diffusion equation is a time dependent parabolic equation where each time level represents an elliptic problem. In these tables, it can also be observed that both of the proposed MEG (AOR) and MEDG (AOR) methods have better execution times than the original EG (AOR) and EDG (AOR) respectively which is due to the reduction in computing efforts of the proposed methods. In the proposed modified group schemes, lesser grid points are involved in the iterative processes than the original group schemes which result in lesser overall arithmetic operation counts

EG(AOR)					MEG(AOR)			
n	R	w	Iter	t	R	w	Iter	t
82	1.308	1.3128-1.3134	13472	13.75	1.109	1.110-1.111	7421	6.766
102	1.3825	1.3843-1.3849	16613	21.875	1.1645	1.1646-1.1656	8703	9.469
122	1.44175	1.44561-1.4458	19841	34.375	1.2134	1.216-1.219	10410	13.453
142	1.49475	1.49903-1.4992	23523	51.172	1.26475	1.2672-1.2676	11672	19.015
162	1.5319	1.5374-1.5375	26931	72.11	1.308	1.3128-1.3134	13472	26.36
182	1.57195	1.5761-1.5762	30634	102.578	1.348	1.3495-1.3499	14914	33.344
202	1.60085	1.60397-1.60402	34307	133.375	1.3825	1.3839-1.385	16614	41.453

Table 1. The numerical performances of the EG(AOR) and MEG(AOR) methods.

EDG(AOR)					MEDG(AOR)			
n	R	w	Iter	t	R	w	Iter	t
82	1.2305	1.238-1.239	10398	9.328	1.066	1.064-1.069	6208	5.89
102	1.302	1.3114-1.3117	12564	14.157	1.105	1.104-1.11	7239	8.89
122	1.367	1.3675-1.3678	14879	19.891	1.15	1.15-1.154	8286	12.391
142	1.4207	1.4213-1.4217	17448	29.406	1.19	1.191-1.195	9343	15.641
162	1.467	1.4694-1.4702	19853	41.235	1.23	1.237-1.24	10396	20.672
182	1.5075	1.50808-1.5082	22344	52.672	1.27	1.2773-1.2786	11480	26.266
202	1.54	1.5449-1.5453	24927	67.266	1.3017	1.311-1.312	12561	32.187

Table 2. The numerical performances of EDG(AOR) and MEDG(AOR) methods

Point C-N (AOR)				
n	R	w	$Iter$	t
82	1.34545	1.35832	15706	25.469
102	1.43985	1.4307	19127	43.906
122	1.4833	1.51066	23884	72.182
142	1.53835	1.5574	27683	101.11
162	1.58092	1.588	30430	148.813
182	1.6301	1.63048	34878	212.141
202	1.66	1.66004	38945	282.328

Table 3. The numerical performance of point *Crank-Nicolson* (C-N) AOR method.

Average Errors					
n	C-N AOR	EG AOR	EDG AOR	MEG AOR	MEDG AOR
82	5.76E-04	2.43E-04	4.13E-04	1.42E-04	5.25E-04
102	5.78E-04	2.31E-04	3.91E-04	1.41E-04	5.16E-04
122	6.35E-04	2.23E-04	3.74E-04	1.38E-04	5.06E-04
142	7.44E-04	2.24E-04	3.64E-04	1.35E-04	4.94E-04
162	8.96E-04	2.35E-04	3.66E-04	1.32E-04	4.81E-04
182	1.07E-03	2.58E-04	3.82E-04	1.29E-04	4.67E-04
202	1.26E-03	2.91E-04	4.11E-04	1.26E-04	4.55E-04

Table 4. Average errors of all the methods for different mesh sizes.

From Tables 1 and 2, it can also be concluded that MEDG (AOR) is able to show the most substantial reduction in execution times compared with the other group and point AOR schemes without having to jeorpadize the solution accuracies. MEDG (AOR) requires the least number of total iterations and computing timings to converge. The required number of iterations is reduced because the introduction of the over-relaxation parameters, w and R, into its formulas is able to reduce the most the number of iterations of the scheme compared with the other schemes tested. This combined with the fact that only about 1/8 of the total nodal points are involved in the iterative process at each time level results in the least computing times for this method. In summary, the proposed MEG (AOR) and MEDG (AOR) methods are viable alternative solvers to the diffusion equation with the latter being the more efficient one in terms of CPU times.

4. Parallel implementations for the group methods

This section will discuss the implementation of the proposed group methods on a message-passing environment.

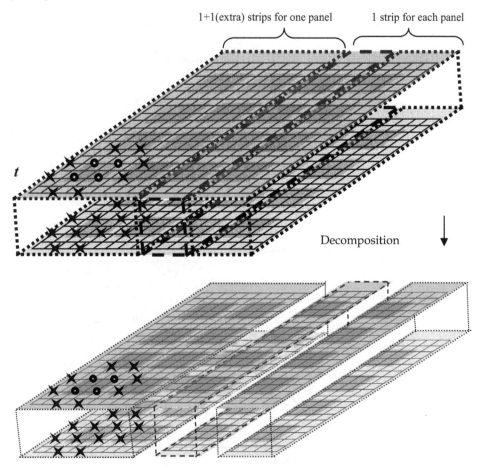

Fig. 6. Domain Decomposition For MEG (AOR) for the case $n = 18$ and $p = 3$

4.1 MEG (AOR) in parallel

For the MEG (AOR) method, we decompose the domain Ω into a number of vertical panels at layer $k+1$ based on the number of available processors, p. The idea is to allocate approximately equal number of strips to the processors. Each strip consists of four grid lines which form the four points blocks with the spacing of $2h$ between the points. The equal number of vertical strips in each panel can be approximated using a specific formula. The distribution of tasks (panels) to processors for the case $n=18$ and $p= 3$ is as shown in Fig. 6 where the configuration is as follows:

- Number of panels = Number of processors, $p = 3$.
- Number of strips in a panel = $(n-2) / 4p = (18-2)/12 = 1$.
- Number of panels that have an extra one strip = $((n-2) \% 4p)/2 = ((18-2) \%12)/4 = 1$.

As shown in Fig. 6, we distribute 1+1(extra strip) strips into panel 1. The other panels (panel 2 and 3) will be allocated with 1 strip of values to update.

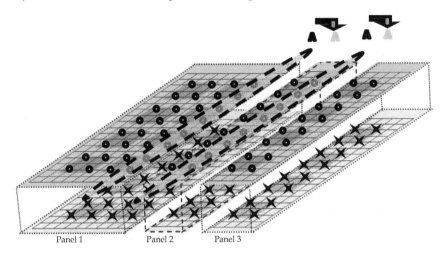

Fig. 7. Send right boundary cell values (grid A's) to left adjacent neighbouring panel.

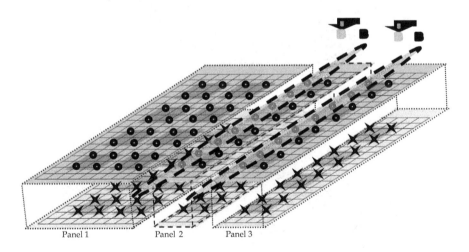

Fig. 8. Send left boundary cell values (grid B's) to right adjacent neighbouring panel.

After the domain Ω is decomposed into the individual panels, message passing needs to be done between the processors to send and receive data at the right and left boundaries of each panel. Based on equations (13)-(14), certain values from adjacent processors need to be communicated during the iterative cycle. The right boundary cell values, grid A's (panels 1 and 2) will be sent to the left adjacent neighbouring panels (panels 2 and 3) as shown in Fig. 7. The left boundary cell values, grid B's (panels 2 and 3) will be sent to the right adjacent neighbouring panel (panels 1 and 2) as shown in Fig. 8. These communications need to be executed correctly to ensure that each processor possesses the correct values needed for their respective independent calculations. After the message passing process is completed, the local error for each processor is calculated and is sent to the master processor for the global convergence check. The local convergence test used is the relative error test similar with their sequential counterparts. The global error is the sum of the local error from each processor. If the global error is greater than a certain tolerance ε, then the iteration is repeated.

4.2 MEDG (AOR) in parallel

Similar with the MEG (AOR) method, we decompose the spatial domain Ω into a number of vertical panels based on the number of available processors, p. For MEDG (AOR), we rotate the x-y axis clockwise 45^0 and forms the four points block with the spacing of $\sqrt{8}h$ between the points of the matrices.

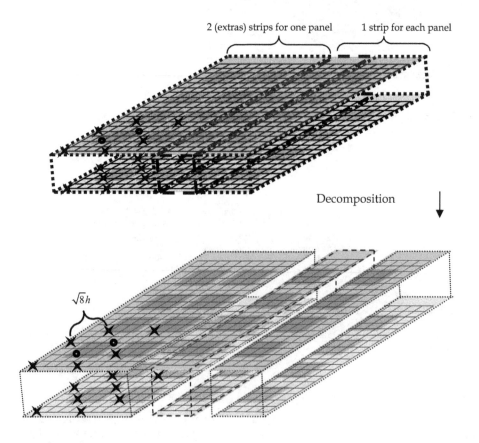

Fig. 9. Domain Decomposition For MEDG (AOR).

We again consider the ordering of the strips for the case $n=18$ and $p=3$ as shown in Fig. 9. Each strip consists of four grid lines which form the four-point groups with spacing $2h$. We will distribute 1+1(extra strip) strips into panel 1. The other panels (panels 2 and 3) will be allocated with 1 strip each to ensure that the tasks are distributed almost equally amongst the processors. The number of strips for each panel (processor) will be computed similar as the one in the previous method.

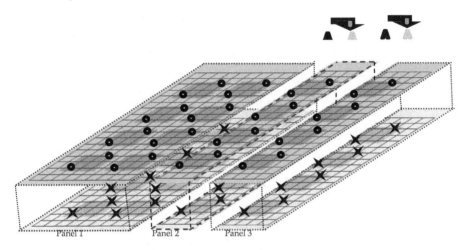

Fig. 10. Send right boundary cell values (grid A's) to the left adjacent neighbouring panel.

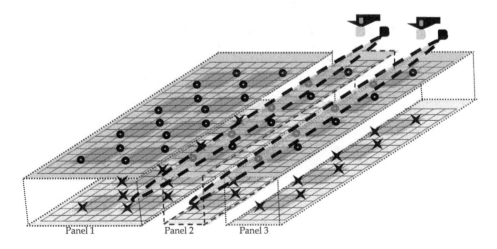

Fig. 11. Send left boundary cell values (grid B's) to the right adjacent neighbouring panel.

After the domain Ω is decomposed into the individual panels, message passing needs to be done between the processors to send and receive data at the right and left boundaries of each panel. The points involved in the iterative process are different from the ones in the previous method due to their different computational molecules. From equations (18)-(19), we can determine these specific values that need to be communicated between adjacent panels during the iterative cycle as shown in Fig. 10 and Fig. 11. The local and global convergences are checked the same way as in MEG.

5. Performance analysis of the parallel group methods

We assume that there are q^2 internal mesh points where $q=n-1$ and arithmetic operations estimates for each method are made per iteration. We assume that the values r, $\Delta tF_{i,j}$, $1-w$, $s_1, s_2, s_3, s_4, s_5, s_6$ are stored beforehand. To update a single block in MEG (AOR) method, the computing cost is given by equation (24):

$$t_{meg(aor)\text{-update}} = 61t_a + 53t_m \tag{24}$$

with t_a = the cost of the addition for a double point and t_m = the cost of the multiplication for a double point. Here, we will consider the problem size, n, and number of processors, p, to have a complete iterative step for the computational cost of MEG (AOR) method which is given by

$$t_{meg(aor)\text{-comp}} = \frac{(q-1)^2}{16p} t_{meg(aor)\text{-update}} \tag{25}$$

The transition cost for message passing of double-type data in a distributed memory multicomputer is given by

$$t_{send} = t_s + qt_d \tag{26}$$

where t_s is the startup time, and t_d is the sending time for a double-type data. The computation of the MEG (AOR) formula requires that q points to be passed to the adjacent processor in an iteration. Therefore, the total communication cost of MEG (AOR) method in a single iterative step, consisting of two sequential point-to-point communications and one global collective communication, is given by

$$t_{meg(aor)\text{-comm}} = 4t_s + 4qt_d \tag{27}$$

After the message passing process is completed, the local error for each processor, p, is calculated and sent to the master processor for the global convergence check. Therefore,

$$t_{meg(aor)\text{-global}} = \frac{(q-1)^2}{4p}(2t_a + t_m) + p(t_s + t_d) \tag{28}$$

The total costs of iterations in parallel MEG (AOR) method is

$$t_{meg(aor)} = t_{meg(aor)\text{-comp}} + t_{meg(aor)\text{-comm}} + t_{meg(aor)\text{-global}} \tag{29}$$

After completing the iteration process, we need to compute at the remaining points by using the *rotated* five points formula for points of type□ and standard 5-points formula for points of type △. This process will be done directly once and the cost of these processes is

$$t_{\text{meg(aor)-once}} = \frac{(q+1)^2}{4} t_{\text{meg(aor)-rotated}} + \frac{(q^2-1)}{2} t_{\text{meg(aor)-standard}} \tag{30}$$

where

$$t_{\text{meg(aor)-rotated}} = 9t_a + 3t_m$$

and

$$t_{\text{meg(aor)-standard}} = 9t_a + 3t_m .$$

For the MEDG (AOR) method, we assume that the values r, $\Delta t F_{i,j}$, $1-w$, s_1, s_2, s_3, s_4 are stored beforehand. To update a single block in this method, the cost is as follows:

$$t_{\text{medg(aor)-update}} = 27t_a + 18t_m \tag{31}$$

with t_a = the cost of the addition for a double point and t_m = the cost of the multiplication for a double point. We will also consider the problem size, n, and number of processors, p, to have a complete iterative step for the computational cost of MEDG (AOR) method which is given by

$$t_{\text{medg(aor)-comp}} = \frac{(q-1)^2}{16p} t_{\text{medg(aor)-update}} \tag{32}$$

The transition cost for message passing of double-type data in a distributed memory multicomputer is given by

$$t_{\text{send}} = t_s + qt_d \tag{33}$$

where t_s is the startup time, and t_d is the sending time for a double-type data. The execution of the MEDG (AOR) formula requires that q points to be passed to the adjacent processor in an iteration. Therefore, the total communication cost of MEDG (AOR) method in a single iterative step, consisting of two sequential point-to-point communications and one global collective communication, is given by

$$t_{\text{medg(aor)-comm}} = 4t_s + 4qt_d \tag{34}$$

After the message passing process is completed, the local error for each processor, p is calculated and is sent to the master processor for the global convergence check. Therefore,

$$t_{\text{medg(aor)-global}} = \frac{(q-1)^2}{8p}(2t_a + t_m) + p(t_s + t_d) \tag{35}$$

As such, the total costs of iteration in MEDG (AOR) method in parallel is

$$t_{\text{medg(aor)}} = t_{\text{medg(aor)-comp}} + t_{\text{medg(aor)-comm}} + t_{\text{medg(aor)-global}} \tag{36}$$

After completing the iteration process, we need to compute the solutions at the remaining points using the standard 5-points formula with the spacing of $2h$ for the points O, rotated 5-points formula for □ and standard 5-points method for △. This process will be done only once directly and the cost of computing these values is

$$t_{medg(aor)\text{-}once} = \frac{(q-1)^2}{8} t_{medg(aor)\text{-}standard\_2\,h} \frac{(q+1)^2}{4} t_{medg(aor)\text{-}rotated} + \frac{(q^2-1)}{2} t_{medg(aor)\text{-}standard} \quad (37)$$

where

$$t_{medg(aor)\text{-}standard\_2\,h} = 9t_a + 3t_m \;,\; t_{medg(aor)\text{-}rotated} = 9t_a + 3\,t_m$$

and

$$t_{medg(aor)\text{-}standard} = 9t_a + 3t_m \;.$$

5.1 Benchmarking

Although it is difficult to obtain reliable estimates for various parameters in any performance models, we run several benchmarking tests on the computing cluster available at the School of Computer Science, Universiti Sains Malaysia (USM), in which the experiments of explicit group methods were carried out. This process is to ensure that we could get the best benchmark with more tests on different time. The specifications of the clusters are shown as below:

a.   Stealth cluster consists of 1 unit of PC with two 900 MHz CPUs, 2GB RAM, and
b.   6 units of PCs each with two 1002 MHz CPUs, 2 GB RAM.
c.   Solaris9 (SunOS 2.9) with Sun HPC ClusterTools 5 and Sun MPI 6.0.

Performance parameter	Benchmark in Stealth cluster
MEG (AOR) point update cost, $t_{meg(aor)\text{-}update}$	1.53 µs/block
MEDG (AOR)point update cost, $t_{medg(aor)\text{-}update}$	1.43 µs/block
sending startup time, t_s	2.3 µs
sending word cost, t_d	0.033 µs/point
global convergence check cost, $t_{meg(aor)\text{-}global}$ and $t_{medg(aor)\text{-}global}$	0.066 µs/point + 2.333 µs/proccessor

Table 5. Performance parameters benchmarking in Stealth Cluster, USM.

5.2 Scalability analysis

By referring to the parameter values in Table 5, we form the performance models for MEG (AOR) and MEDG (AOR) methods which are shown as the following:

a)   MEG (AOR): $t_{method}(n,p) = A\dfrac{(q-1)^2}{16p} + B + Cq + D\dfrac{(q-1)^2}{4p} + Ep$,   (38)

b)   MEDG (AOR): $t_{method}(n,p) = A\dfrac{(q-1)^2}{16p} + B + Cq + D\dfrac{(q-1)^2}{8p} + Ep$ (39)

where A, B, C, D and E are coefficients of the methods which shown in Table 6.

Method	A	B	C	D	E
MEG (AOR)	1.53	9.2	0.132	0.066 μs	2.333
MEDG (AOR)	1.43	9.2	0.132	0.066 μs	2.333

Table 6. Coefficients of the performance models in μs.

6. Numerical results of the parallel group accelerated methods

We implement the parallel algorithms on the Stealth cluster at USM. The experiments were carried out on 1 unit of PC with two 900 MHz CPUs, 2 GB RAM, and 6 units of PC, where each PC had two 1002 MHz CPUs and 2 GB RAM. The Operating System used was Solaris9 (SunOS 2.9) with Sun HPC ClusterTools 5 and Sun MPI 6.0. The parallel algorithms were tested on the same model problem that was used for the sequential version (23). For the MEG (AOR) and MEDG (AOR) methods, the sizes of n were chosen appropriately to make sure that all of the strips consisting of two grid lines can be decomposed approximately evenly to the 6 processors. The tolerance used was $\varepsilon = 1.0 \times 10^{-6}$ and the acceleration parameters, w and R, were chosen to give the least number of iterations.

From Table 6, we can see that the computation coefficient of MEDG (AOR) is slightly lesser than MEG (AOR). Therefore we expect that MEDG (AOR) should have better timings if compared to MEG (AOR). We further test the scalability analysis by comparing the experimental and predicted timings of these methods using $n = 182$ and 202 which are shown in Figs. 12 and 13 respectively.

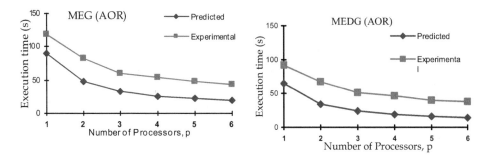

Fig. 12. Comparison of Predicted Timings and Experimental Timings of parallel MEG (AOR) and MEDG (AOR) methods for n = 182.

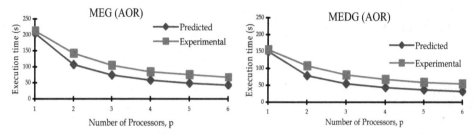

Fig. 13. Comparison of Predicted Timings and Experimental Timings of parallel MEG (AOR) and MEDG (AOR) methods for n = 202.

The figures show that the experimental and predicted timings are very close to one another especially when the grid size is larger. Comparing between the two grid sizes, it is found that the efficiency improves as the grid size increases. This improvement indicates that the performance models are more accurate as the grid sizes increase. Based on the parallel implementation which was described in Section 4, we used the size of $n = 162, 182$ and 202 to record the timings, speedups and efficiencies of both the MEG (AOR) and MEDG (AOR) methods. Several performance results of the MEG (AOR) and MEDG (AOR) methods are shown in Figs. 14-16.

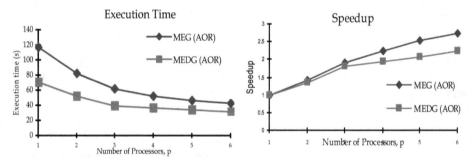

Fig. 14. Comparisons of execution time (Left) and speedup values (Right) between MEG(AOR) and MEDG(AOR) methods for $n = 162$.

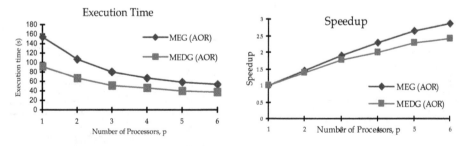

Fig. 15. Comparisons of execution timings (Left) and speedup values (Right) between parallel MEG(AOR) and MEDG(AOR) methods for $n = 182$.

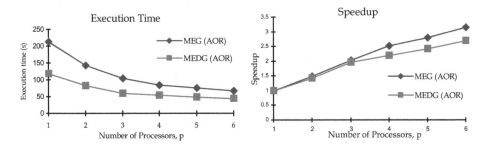

Fig. 16. Comparisons of execution timings (Left) and speedup values (Right) between parallel MEG(AOR) and MEDG(AOR) methods for $n = 202$.

From these figures, we can see that the parallel MEDG (AOR) is better in execution timings compared to the MEG (AOR). Generally we can see that with the enhancement of grid size, the speedup increases with nearly 70% efficiency. However, the speedup and efficiency values of MEG (AOR) are slightly better than MEDG (AOR). This difference in values indicates that the amount of computations carried out over the total communication overheads in MEG(AOR) is greater than the one in MEDG(AOR).

7. Conclusions

In this chapter, the formulation of new improved explicit group AOR methods in solving the two dimensional diffusion equation is presented. The improvement of the numerical result shows the potential of these methods in solving the parabolic equation. We further implement both of these methods on a cluster of distributed memory computer using Message-Passing Interface programming environment. The experimental results show that these two methods can be performed successfully in parallel on a cluster of distributed memory computer. Performance models to explain the parallel behaviour of these proposed methods were also developed. The experimental timings agreed with the predicted results especially when the grid size and processors increase. The MEDG (AOR) shows a faster rate of convergence with similar accuracies if compared with MEG (AOR), especially when the grid size increases. Both methods were shown to be suitable to be programmed on a distributed memory computer.

8. Acknowledgement

The authors acknowledge the Fundamental Research Grant Scheme (203/ PMATHS/ 6711188) for the completion of this work.

9. References

Abdullah, A. R. (1991). The Four Point Explicit Decoupled Group (EDG) Method: A Fast Poisson Solver. *International Journal of Computer Mathematics.* Vol. 38, pp. 61-70

Ali, N. H. M. & Lee, S. C. (2007). Group Accelerated Over Relaxation Methods On Rotated Grid. *Applied Mathematics and Computation*, Vol. 191, pp. 533-542

Ali, N,H.M. (1998). *The Design And Analysis of Some Parallel Algorithms For The Iterative Solution of Partial Differential Equations*, PhD Thesis, Universiti Kebangsaan Malaysia.

Evans, D.J. & Abdullah, A.R. (1983a). A New Explicit Method for the Solution of . *International Journal of Computer Mathematics*, Vol. 14, pp. 325-353

Evans, D.J. & Abdullah, A.R. (1983b). Group Explicit Methods for Parabolic Equations. *International Journal of Computer Mathematics*. Vol. 14, no. 1, pp. 73-105

Evans, D.J. & Sahimi, M. S. (1988). The Alternating Group Explicit(AGE) Iterative Method for Solving Parabolic Equations, 1-2 Dimensional Problems. *International Journal of Computer Mathematics*, Vol. 24, pp. 250-281

Hadjidimos, A. (1978). Accelerated OverRelaxation method. *Mathematics of Computation*, Vol. 32, pp. 149–157

Martins, M.M., Yousif, W.S. & Evans, D.J. (2002). Explicit group AOR method for solving elliptic partial differential equations. *Neural, Parallel and Science Computation*, Vol. 10, no. 4, pp. 411-422

Ng, K. F. & Ali, N. H. M. (2008). Performance Analysis of Explicit Group Parallel Algorithms for Distributed Memory Multicomputer. *Parallel Computing*, Vol. 34, no (6-8), pp. 427-440

Othman, M. & Abdullah, A.R. (2000). An Efficient Four Points Modified Explicit Group Poisson Solver. *International Journal of Computer Mathematics*, Vol. 76, pp. 203-217.

Yousif, W.S. & Evans, D.J. (1986). Explicit Group Over-relaxation Methods for Solving Elliptic Partial Differential Equations. *Mathematics & Computers In Simulation*, Vol. 28, pp. 453-466.

4

An Idealised Biphasic Poroelastic Finite Element Model of a Tibial Fracture

Sanjay Mishra

School of Engineering Systems, Queensland University of Technology, Brisbane, Australia

1. Introduction

The outcome of a bone fracture partly depends upon the mechanical environment experienced by the fracture callus (reparative tissue) during the healing. Therefore biomechanics of bone fracture healing has been examined in many clinical or biological, mathematical or finite element studies (Cheal et al. 1991, DiGioia et al. 1986, Claes et al. 1999, Doblaré et al. 2004 and Oh et al. 2010). Most of the studies model the components of bone fractures as monophasic, homogenous materials, which may not be appropriate considering the large inter fragmentary displacements and high porosity of the reparative tissue. Therefore, this study describes an idealised mathematical model of a healing bone fracture with biphasic approach when the callus bone is modelled as mixture of solids and fluids.

Markel *et al.* (1990) reported that the porosity of the callus in a healing canine osteotomy decreased from 99.6% at 2 weeks to 38% at 12 weeks. Therefore, the biphasic, poroelastic model for fracture callus and bone has been suggested in the literature (Carter *et al.* 1998, Simon *et al.* 1992, Prendergast et al1997, Spilker *et al.* 1990). Biphasic poroelastic models for soft tissues (Mow *et al.* 1980, Simon *et al.* 1985, Van Driel *et al.* 1998, Prendergast *et al.* 1997, Spilker *et al.* 1990) have been developed and applied to model cartilage (Mow *et al.* 1980) and intervertebral discs (Simon *et al.* 1985). Van Driel *et al.* (1998) and Prendergast *et al.* (1997) modelled tissue adjacent to prostheses using poroelastic material properties to investigate tissue differentiation. In the field of fracture healing however, only monophasic material properties of callus have been simulated (Carter 1988, Carter 1998, Blenman 1989, Cheal 1991, DiGioia 1986, Claes 1999, Gardner 1998 and 2000). This is probably because of the paucity of data in the literature on the values of parameters required to define the biphasic material properties of fracture callus. Simulation of a biphasic, compressible, anisotropic, linear poroelastic material model requires forty material constants (Simon 1992). Even the very simplified simulation of an isotropic material requires a minimum of five material constants. However, the number of material constants required to simulate a biphasic, poroelastic medium can be further reduced to three if the solid and fluid media are assumed to be incompressible (Simon 1992). These three independent material parameters are Lame's material stiffness parameters (λ and μ) and hydraulic permeability (k). Alternatively, Zienkiewicz and Taylor (1994b) suggested a method to model poroelastic behaviour under `undrained' condition using the modulus of elasticity, Poisson's ratio, the porosity of the matrix, and the bulk modulus of the fluid phase. In the present study, a finite element model (FEM) based on the poroelastic behaviour of the 'undrained' callus at four temporal stages of healing is developed by modifying the theory proposed by Zienkiewicz and Taylor (1994). This model was developed to examine the influence of fluid pressure on the pattern of healing and to compare the distribution of stresses in the callus with the monophasic solutions developed for the same subject at the same temporal points reported earlier by Gardner *et al.* (2000).

2. Materials and methods

2D, monophasic, plane stress, FEM's of a mid-diaphyseal tibial fracture were developed at four stages during healing (4, 8, 12, and 16 weeks post operation) by Gardner *et al.* (2000). The geometry and the regionalisation of the callus are shown in Figure 1. The geometry, finite element meshing, boundary conditions (Figure 2) and applied displacements (Table 1) used in the study of Gardner *et al.* (2000) are adopted in the biphasic poroelastic models of the present study. The tissue histology and calculated elastic moduli of regions of callus at four stages of healing are shown in Table 2.

Week	x(mm)	y(mm)	Z(radians)
4	.283	-.377	-.00329
8	-.015	-.159	.00201
12	-.092	-.260	.00169
16	-.129	-.111	-.00215

Table 1. Interfragmentary displacements measured during walking corresponding to peak longitudinal displacements. y (longitudinal), x (transverse), Z (rotational in x-y plane) adopted from the study of Gardner *et al.* (2000).

Stage	CENTRAL		ADJACENT		PERIPHERAL	
weeks	tissue type	Modulus (MPa)	tissue type	Modulus (MPa)	tissue type	Modulus (MPa)
4	Haematoma, granulated tissue	0.9	Soft connective tissue with invading vasculature	3.8	Soft fibrocartilage tissue	76
8	Fibrous perichondrial tissue	28	Woven bone, 25% maturation	700	Dense fibrous tissue, 45% maturation	2800
12	Fibrous perichondrial tissue	30.6	Woven bone, 25% maturation	765	Dense fibrous tissue, 45% maturation	3060
16	Fibrous cartilage tissue, 10% maturation	75	Woven bone, 60% maturation	5000	Bone, 100% maturation	20000

Table 2. Tissue histology and calculated Young's moduli of the three regions of fracture callus, at 4, 8, 12 and 16 weeks post fracture adopted from the study of Gardner *et al.* (2000).

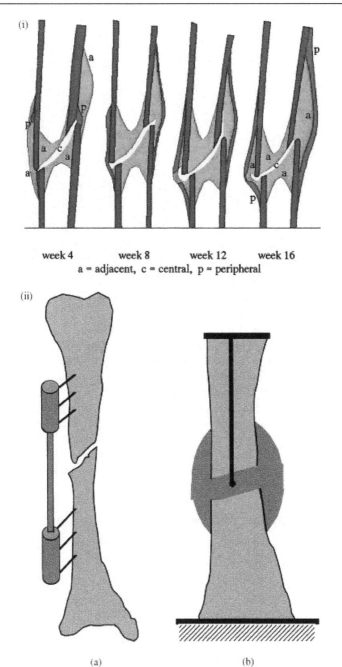

Fig. 1. (i) Regions of cortical bone and callus in the 2D finite element models a- adjacent (green), c- central (yellow), p- peripheral (red), (ii) Schematic diagram showing the bone fracture, external fixation device and callus region (adopted from Gardner *et al. (2000)*.

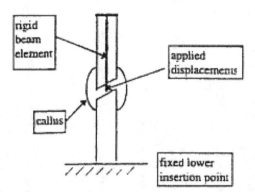

Fig. 2. Boundary condition of the finite element model, showing the fixed lower boundary and the displacement of the upper bone fragment applied at the fracture centre adopted from the Gardner *et al.* (*2000*).

However, the callus tissue in the present study was idealised as a homogenous, fully saturated, linear poroelastic medium consisting of a matrix of solid and incompressible fluid, as opposed to the monophasic material properties in the study of Gardner *et al.* (2000). The following section describes the theory employed for calculating equivalent poroelastic material properties of the callus from the modulus of elasticity, Poisson's ratio, porosity and bulk modulus of the fluid of the callus tissue. Unlike a monophasic medium, the normal stress acting across a plane within a biphasic poroelastic mass will have two components, an inter-granular pressure known as effective pressure or effective stress, and a fluid pressure called the pore pressure. The sum of these two will constitute the total normal stress. The volume change characteristics and the strength of poroelastic mediums are controlled by the effective stress not by the total stress. Thus the only difference between the present study and the previous study of Gardner et al. (2000) is that the constitutive equation of the callus is changed from monophasic medium to a biphasic poroelastic medium.

3. Mathematical description of the model

For 2D plane stress analysis, the constitutive equation for a monophasic material (Zienkiewicz and Taylor (1994a)) is

$$\sigma = \left\{ \begin{array}{c} \sigma_x \\ \sigma_y \\ \tau_{xy} \end{array} \right\} = \mathbf{D} \left\{ \begin{array}{c} \varepsilon_x \\ \varepsilon_y \\ \gamma_{xy} \end{array} \right\} \tag{1}$$

Where σ is total stress. σ_x , σ_y are normal stresses in the x and y directions, τ_{xy} is shear stress in the x-y plane, ε_x, ε_y are normal strains in the x and y directions and γ_{xy} is the shear strain in the x-y plane. For plane stress analysis the constitutive matrix $\mathbf{D}$ (Zienkiewicz and Taylor 1994a) is defined as

$$D = \frac{E}{1-v^2}\begin{bmatrix} 1 & v & 0 \\ v & 1 & 0 \\ 0 & 0 & \dfrac{1-v}{2} \end{bmatrix} \tag{2}$$

Where E and v are the Young's Modulus and the Poisson's ratio of the material.
A plane stress, linear elastic finite element program can be used to analyses the linear elastic plane strain problem (Zienkiewicz and Taylor (1994a)) by substituting

$$E = E_s = \frac{E_n}{1-v_n^2} \tag{3a}$$

and

$$v = v_s = \frac{v_n}{1-v_n} \tag{3b}$$

Where subscript 's' denotes plane stress parameters and 'n' denotes plane strain parameters. Substituting equation (3a) and (3b) into equation (2), we have

$$D = \frac{\dfrac{E_n}{1-v_n^2}}{1-\left(\dfrac{v_n}{1-v_n}\right)^2}\begin{bmatrix} 1 & \dfrac{v_n}{1-v_n} & 0 \\ \dfrac{v_n}{1-v_n} & 1 & 0 \\ 0 & 0 & \dfrac{1-\dfrac{v_n}{1-v_n}}{2} \end{bmatrix} \tag{4a}$$

Simplifying

$$D = \frac{E_n(1-v_n)^2}{(1-2v_n)(1-v_n^2)}\begin{bmatrix} 1 & \dfrac{v_n}{1-v_n} & 0 \\ \dfrac{v_n}{1-v_n} & 1 & 0 \\ 0 & 0 & \dfrac{1-2v_n}{2(1-v_n)} \end{bmatrix} \tag{4b}$$

$$D = \frac{E_n(1-v_n)}{(1-2v_n)(1+v_n)}\begin{bmatrix} 1 & \dfrac{v_n}{1-v_n} & 0 \\ \dfrac{v_n}{1-v_n} & 1 & 0 \\ 0 & 0 & \dfrac{1-2v_n}{2(1-v_n)} \end{bmatrix} \tag{4c}$$

Thus a linear elastic plane strain **D**-matrix is obtained.
The equation for static equilibrium in 2D (Dawe 1984) is:

$$(\partial \sigma_x / \partial x) + (\partial \tau_{xy} / \partial y) + R_x = 0 \tag{5a}$$

$$(\partial \sigma_y / \partial y) + (\partial \tau_{xy} / \partial x) + R_y = 0 \tag{5b}$$

Using the definition of effective stress (Zienkiewicz and Taylor (1994b))

$$\sigma_X = \sigma'_X - p \tag{6a}$$

$$\sigma_Y = \sigma'_Y - p \tag{6b}$$

$$\tau_{XY} = \tau'_{XY} \tag{6c}$$

where $\sigma_X, \sigma_Y, \tau_{XY}$ are total stresses; R_x, R_y are body forces; σ'_X, σ'_Y, τ'_{XY} are effective stresses (positive value for tensile stresses and negative for compressive stresses), p is pore pressure of the fluid, which is conventionally described as positive for compressive pressure and negative for tensile pressure.

The combined seepage and conservation of fluid equation described by Zienkiewicz and Taylor (1994b) is:

$$-\frac{\partial}{\partial x}\left(k\frac{\partial p}{\partial x}\right) - \frac{\partial}{\partial y}\left(k\frac{\partial p}{\partial y}\right) + \frac{\dot{p}}{Q} + \dot{\varepsilon}_X + \dot{\varepsilon}_Y = 0 \tag{7}$$

Where k is the permeability of the fluid, Q is the ratio of the bulk modulus of fluid to the porosity of the media and ε_X, ε_Y represent the volumetric strain rates of the solid skeleton. The superscript 'dot' denotes differentiation with respect to time. Since the gait cycle frequency of this clinical fracture was approximately 1 Hz, it is reasonable to assume that little or no seepage from the callus occurs during loading (Carter 1998, Gardner 1998). Therefore, because the time scale is short, if the local `undrained' condition of the callus is assumed, permeability 'k' can also be assumed to be effectively zero. Under these conditions, Equation 7 becomes:

$$\frac{\dot{p}}{Q} + \dot{\varepsilon}_X + \dot{\varepsilon}_Y = 0 \tag{8}$$

Or

$$\dot{p} = -Q(\dot{\varepsilon}_X + \dot{\varepsilon}_Y) \tag{9}$$

Integrating with respect to time and assuming homogenous initial conditions ($p = 0$, $\varepsilon_X = \varepsilon_Y = 0$ at $t = 0$):

$$p = -Q(\varepsilon_X + \varepsilon_Y) \tag{10}$$

Substituting this value in equation (6a) and (6b) gives:

$$\sigma_{xx} = \sigma'_{xx} + Q(\varepsilon_X + \varepsilon_Y) \tag{11}$$

$$\sigma_{yy} = \sigma'_{yy} + Q(\varepsilon_X + \varepsilon_Y) \tag{12}$$

Expanding for two-dimensional plane strain analysis ($\sigma_z \neq 0$) gives:

$$\begin{Bmatrix} \sigma_x \\ \sigma_y \\ \sigma_z \\ \tau_{xy} \end{Bmatrix} = \begin{Bmatrix} \sigma'_x \\ \sigma'_y \\ \sigma'_z \\ \tau'_{xy} \end{Bmatrix} + Q \begin{Bmatrix} \varepsilon_x + \varepsilon_y + \varepsilon_z \\ \varepsilon_x + \varepsilon_y + \varepsilon_z \\ \varepsilon_x + \varepsilon_y + \varepsilon_z \\ 0 \end{Bmatrix} \tag{13}$$

Noting that the material behaviour is controlled by the effective stress in a poroelastic medium, then

$$\sigma' = D\varepsilon \tag{14}$$

Combining equation (13) and (14):

$$\begin{Bmatrix} \sigma_x \\ \sigma_y \\ \sigma_z \\ \tau_{xy} \end{Bmatrix} = D \begin{Bmatrix} \varepsilon_x \\ \varepsilon_y \\ \varepsilon_z \\ \gamma_{xy} \end{Bmatrix} + \begin{bmatrix} Q & Q & Q & 0 \\ Q & Q & Q & 0 \\ Q & Q & Q & 0 \\ 0 & 0 & 0 & 0 \end{bmatrix} \begin{Bmatrix} \varepsilon_x \\ \varepsilon_y \\ \varepsilon_z \\ \gamma_{xy} \end{Bmatrix} \tag{15}$$

If the modified stress-strain relationship is defined as:

$$\sigma = D'\varepsilon \tag{16}$$

then

$$D' = \begin{bmatrix} D_{11}+Q & D_{12}+Q & D_{13}+Q & 0 \\ D_{21}+Q & D_{22}+Q & D_{23}+Q & 0 \\ D_{31}+Q & D_{32}+Q & D_{33}+Q & 0 \\ 0 & 0 & 0 & D_{44} \end{bmatrix} \begin{Bmatrix} \varepsilon_x \\ \varepsilon_y \\ \varepsilon_z \\ \gamma_{xy} \end{Bmatrix} \tag{17}$$

and

$$Q = K_f / n, \tag{18}$$

where K_f is the bulk modulus of the fluid phase and n is the porosity of the porous media. Thus the **D'** matrix (Equation 17) for a poroelastic medium can be calculated by substituting Q values in the **D** matrix of corresponding plane strain analysis (Equation 4C). The **D** matrix can be expanded to calculate the new set of material properties, E and ν, corresponding to the poroelastic material behaviour under undrained condition.

4. Development of the poroelastic FEM from the monophasic FEM

Using the above theory, a new set of material properties (E and ν) were calculated from the refined values of E and ν of the monophasic model of Gardner et al. (2000) and the values of Q described in this section. At first, the proportion of calcified tissue in the callus was extrapolated from data (at $E = 800$ MPa 20% calcification; at $E = 2000$ MPa 40% calcification;

at E = 8000 MPa 70% calcification and at E = 18000 MPa 100% calcification occurs) taken from Davy and Connoly (1982) for the new bone at intermediate densities corresponding to the elastic moduli of the callus adopted from the study of Gardner *et al.* (*2000*). The calculated calcification of the different regions of callus at 4, 8, 12 and 16 weeks is shown in Table 3. The porosity of the callus tissue was then assumed to be inversely related to the proportion of the calcified tissue present in the callus (Carter 1977), and therefore was found to vary from 0.9 for soft callus (<5% calcification) to 0.3 for woven bone (100% calcification). As the porosity of 0.8 was used by Van Driel *et al.* (1998) for soft fibrous, cartilage and bone tissues, by comparison 0.9 appeared valid for the softer tissues of the present model. The porosity of 0.3 was suggested for woven bone (Carter 1977) which seems to be valid for the harder tissues of the present model. All intermediate values of porosities were linearly extrapolated from these two values at corresponding values of calcification shown in Table 3. The interstitial fluid in the callus was assumed to have the bulk modulus (K_f) of salt water (2.3 GPa) (Cowin 1999).

Time weeks	Callus	E_m (MPa)	v_m	Calcification %	n	Q (K_f/n)	E_p (MPa)	v_p
4	central	0.9	0.39	<5	0.9	2555	0.2	0.499
	adjacent	3.8	0.39	<5	0.9	2555	4.38	0.499
	Peripheral	76	0.39	8.8	0.88	2643	81	0.494
8	central	28	0.39	5	0.89	2613	30	0.498
	adjacent	700	0.30	24	0.78	2948	786	0.46
	Peripheral	2800	0.30	44	0.64	3593	3023	0.43
12	central	30.6	0.39	5	0.89	2613	33	0.497
	adjacent	765	0.30	25	0.78	2948	858	0.457
	Peripheral	3060	0.30	46	0.64	3593	3291	0.398
16	central	75	0.30	8.8	0.88	2643	86	0.494
	adjacent	5000	0.30	57	0.57	4035	5283	0.373
	Peripheral	20000	0.30	100	0.3	7666	20381	0.324

Table 3. Calculation of the equivalent poroelastic material properties of the callus from the monophasic material properties. Note: subscript 'm' denotes the monophasic model and 'p' denotes the biphasic poroelastic model. K_f = 2.3 GPa (Cowin 1999).

5. Results

Figures 3 and 4 show the contour diagrams of the fluid pressure, effective stress and total stress in the callus at 4, 8, 12 and 16 weeks post fracture. Additional figures of effective stresses showing only one region of the callus were also drawn for clarity and the results are shown in Table 4. At Week 4 (Figure 3a) the peak compressive fluid pressures (≥ 400 MPa) were present in the cortical gap region, intermediate pressures (≥ 150 MPa) were present in the other regions of the interfragmentary gap and the periosteal regions close to the interfragmentary gap, and low pressures (≤ 50 MPa) were present in the regions of the callus remote to the interfragmentary gap. Tensile pressures of up to 100 MPa were present in localised regions of the endosteal callus. At Week 8 [Figure 3(d)] fluid pressures in all regions of the callus were reduced (< ± 5 MPa) but elevated pressures (>70 MPa) were present in the cortical gap regions. At Week 12 [Figure 4(a)], fluid pressures of greater than 100 MPa are seen medially and less than 25 MPa laterally in the cortical gap region. All other regions indicated low pressures (<10 MPa); however, tensile pressures of about 30 MPa are indicated in the subperiosteal region on the lateral side. At Week 16 [Figure 4 (d)], fluid pressures are further reduced in all regions of the callus, although in the medial inter-cortical gap they remained elevated (15 < P < 25 MPa). Inter cortical fluid pressures are reduced laterally almost to the level of the periosteal callus (0 < P < 1 MPa).

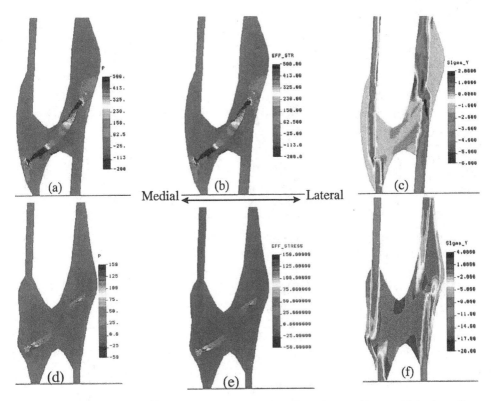

Fig. 3. Fluid pressure (p), effective stress (EFF_STR) and total stress (Sigma_Y) in the callus at 4 weeks [(a), (b) and (c)] and at 8 weeks [(d), (e) and (f)].

Effective stress diagrams are similar in magnitude and distribution to the fluid pressure diagrams. As expected, the total stresses (σ_Y) were very low as compared to the corresponding fluid pressures (*p*) because during undrained loading condition, most of the load is taken by fluid medium. Also, in the regions of low fluid pressures the effective stress magnitudes (Table 4) are approximately equal to the total stresses. The total pressure diagrams [Figures 3 and 4 (c) (f)] are similar to the corresponding longitudinal stress diagrams of the monophasic model reported earlier by Gardner *et al.* (*2000*).

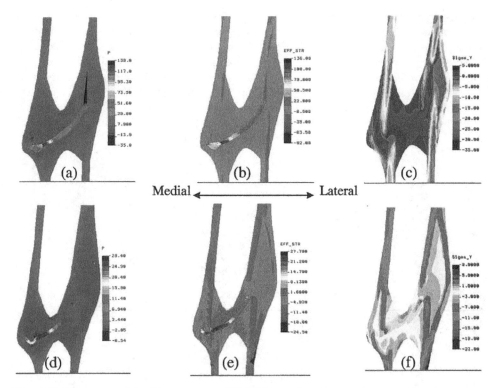

Fig. 4. Fluid pressure (p), effective stress (EFF_STR) and total stress (Sigma_Y) in the callus at 12 weeks [(a), (b) and (c)] and at 16 weeks [(d), (e) and (f)].

Time in Weeks	Callus region	Monophasic model Longitudinal stress (MPa)	Biphasic model Effective stress (MPa)
4	Central	-1.5 to 0.5	100 to 1000
	Adjacent	-1.5 to 0.5	100 to 500
	Peripheral	-1.5 to 0.5	25 to 500
8	Central	-1 to 2	25 to 150
	Adjacent	-5 to 2	-1 to 1
	Peripheral	-15 to -1	-1 to 1
12	Central	-3 to 1	100 to 500
	Adjacent	-8 to 1	25 to 50
	Peripheral	-3 to -15	0 to 12
16	Central	0 to 4	100 to 150
	Adjacent	-3 to 0	-1 to 1
	Peripheral	-3 to -11	-1 to -5

Table 4. Longitudinal stress ranges in the monophasic model of Gardner *et al.* (*2000*) and effective stress ranges in the biphasic poroelastic model.

6. Discussion and conclusions

The magnitudes of fluid pressure and effective stress are very high and therefore appear unrealistic. In particular, the high tensile stress would produce cavitation or may lead to gas in the pore fluid, and the callus matrix may be ruptured. These conditions are typical of the undrained simulation behaviour under large deformations. In reality, no matter how small the permeability of the poroelastic medium or how rapid is the loading, fluid flow will occur under such high-pressure gradient. Therefore the absolute value of pressure and the presence of high tensile pressure in the callus are more an artifact of the modelling technique than are the patterns of the distribution of pressure and their trend in variation. Thus, for this technique of modelling the presence of spatial or temporal pressure gradients within the callus may be a valid indicator of the flow of fluid in the form of blood, nutrients or waste products. For example, the tensile fluid pressure regions of the present models at peak loading during walking are likely to show reduced magnitudes of tensile pressure or compressive pressure during unload phases. The hydraulic gradient will be reversed and the fluid flow will be in the opposite direction. This alternate inflow and outflow of fluid may be related to the transport mechanism of inflow of nutrients and oxygen, and outflow of waste products and carbon dioxide from the callus. Such inflow and outflow may enhance the growth of capillary blood vessels, thus accelerating the healing process.

It can be envisaged from the above that if the movement is too small, the change in fluid pressure will be small and the beneficial effect will also be small. Therefore the results of the present study corroborate those of other studies (Kenwright 1998, Sarmiento and Latta 1995, Goodship and Kenwright 1985, Kenwright and Goodship 1989) suggesting that fracture site movement is necessary for efficient secondary healing. However, if the movement is too large the fabric of the callus matrix could be damaged because of the cyclical expansion and compression, and this damage could hinder healing.

Furthermore, if the frequency of the movement were too high, the time interval between pressure gradient reversals would be too small for the fluid to flow. On the other hand, if the frequency is too low, fluid will penetrate the tissues before any substantial hydraulic gradient can be developed. Since the present study was limited to a single temporal point during the gait cycle it is not possible to define the optimum magnitude and frequency of movement beneficial for healing. Models that simulate the different temporal points of the gait cycle may provide more information about the optimum movements for fracture patients.

At Week 4 (Figure 3a), there are high compressive fluid pressures in the inter fragmentary gap regions because the undrained model is under large compressive displacement and fluid is unable to flow out side the system boundary. In this condition the loads is predominantly taken by an incompressible fluid that controls the motion of the bone fragments and provides support for intact tissues. It does this by increasing the stiffness of the limb and it also protects the fracture from further damage (Sarmiento and Latta 1995). Since fluid pressure is a function of movement, the cortical gap locations are expected to undergo higher pressures than locations further away from the gap, as shown in Figure 3a. It is worth noting that at this stage the callus is comprised of more than 90% fluid, therefore the incompressible fluid will resist high pore pressure.

At Week 8, reduction in the porosity (Table 3) and compressive interfragmentary displacement (Table 1) of the callus reduce fluid pressure. However, at Week 12, the porosity remains almost constant (Table 3) compared to Week 8 but the applied compressive interfragmentary displacement increases (Table 1), resulting in elevated fluid pressures in the adjacent callus. At Week 16, both the applied compressive interfragmentary displacement and the fluid pressure in the callus reduce. Therefore fluid pressures appear to be more sensitive to longitudinal interfragmentary displacements than to the callus-porosity in the present study.

Fluid pressure distribution patterns correlate with the general pattern of ossification, as reported by others (Blenman 1989,Carter 1988, Sarmiento 1995, Yamagishi 1955). Blenman and Carter (1989) suggested that ossification progresses through the stages of `bone tuft', `bone wedge' and `bone bridge' and poroelastic models appear to corroborate this if it is believed that ossification may not take place in the regions of high fluid pressures. At Week 4, high pore pressure regions of the interfragmentary gap divide the proximal and distal callus. Therefore it appears that ossification is possible only in the low-pressure regions of the periosteal callus away from the interfragmentary gap, forming the `wedge' shaped ossified callus. At Weeks 8 and 12, fluid pressure in the callus is reduced almost to zero at the level of the gap, thus allowing the formation of a `bridge' of ossified callus between the `wedges'. Since the present study started at 4 weeks post operation, a pattern similar to that of the `tufts' theory may have also been present before this stage.

The similarity between the magnitude and pattern of total stress diagrams of the biphasic poroelastic model and the corresponding longitudinal stress diagrams of the monophasic model of Gardner et al. (2000) are expected. This is because total stress is a function of the total force and total cross section area (Wood 1990) of the callus, which remain similar in both the models.

The greatest disparities between the monophasic and biphasic solutions occur in initial healing at Week 4. This disparity exists because the soft callus has a high porosity initially and is subjected to high tensile stresses during the large applied initial displacements. As the callus calcifies, its porosity and the fluid pressures decrease so that total stress is closer to the effective stress. This effect of porosity is evident from Table 4, where the maximum difference between stresses from the two models is in the high-porosity central callus and the minimum difference is in the low-porosity peripheral callus. The patterns of effective stress in the biphasic models also differ from the corresponding monophasic models. In the biphasic models, substantial variations of stress are evident at the cortical gap, sub periosteal and endosteal callus. Whereas in general, the monophasic models reported by Gardner et al. (2000) predicts similar stress regimes throughout the central and adjacent callus. Therefore, if the differentiation and maturation of the callus are believed to be influenced by the preceding stress environments, then the biphasic models appear to predict more realistic patterns of tissue differentiation and maturation.

7. Limitations of the study

The results of the present study should be evaluated under modelling limitations. Firstly, solutions are valid only for the `undrained' condition that assumes that no fluid moves out of the system boundary during loading, but in reality a small amount of fluid may drain through the pores of the callus. However this drainage may not have invalidated the results of the present study as the gait cycle frequency is approximately 1 Hz, and physiological loading periods during the stance phase of gait are around 0.3 to 0.5 seconds, which is probably too rapid for significant drainage of fluid to take place (Gardner et al. 1998, *2000*). Secondly, the callus has been idealised as a linear, elastic, fully-saturated porous medium throughout healing. These idealised conditions may also be responsible for the high magnitudes of fluid pressure, whereas actual pressures are believed to be lower than predicted by the present model. However, the pattern and trend of temporal variations in fluid pressures during progressive ossification are unlikely to change significantly as a result of applying slightly different poroelastic material properties and constitutive equations.

Despite the limitations of the present model, these results indicate that the biphasic material properties of the callus are more appropriate to the initial soft callus stage of healing and support the suggestion of Sarmiento and Latta (1995) "The incompressible fluid effect, or hydraulics is most important in early post injury period. We feel that hydraulics is responsible for the control of motion of fragments before callus has developed and that it provides the significant degree of stiffness observed in loaded limbs with fresh fractures fit with fracture braces."

8. Acknowledgement

The author wishes to acknowledge (a) the financial support obtained by commonwealth Commission UK during the study (b) the major input and supervision from Prof AHC Chan and Dr Trevor Gardner, University of Birmingham UK.

9. References

Blenman, P. R., Carter, D. R., and Beaupre, G. S. (1989) Role of mechanical loading in the progressive ossification of a fracture callus. *J. Orthop. Res.* 7, 398-407.

Carter D, Hayes WC. (1977) The compressive behaviour of bone as a two phase porous structure. JBJS,59A (7):954-962

Carter D. R., Blenman P. R., and Beaupre G. S. (1988) Correlations between mechanical stress history and tissue differentiation in initial fracture healing. *J. Orthop. Res.* 6, 736-748.

Carter D. R., Beaupre G.S., Giori N.J., and Helms J.A. (1998) Mechanobiology of skeletal regeneration. Clin. Orthp. Rel. Res., 355 S: 41-55

Carter D. R., Blenman P. R., and Beaupre G. S. (1988) Correlations between mechanical stress history and tissue differentiation in initial fracture healing. *J. Orthop. Res.* 6, 736-748.

Dawe D.J., (1984) Matrix and finite element displacement analysis of structures. Clarendon press, Oxford (UK).

Doblaré M , García J.M., and Gómez J.M., (2004). Modelling bone tissue fracture and healing: a review; Engineering Fracture Mechanics, 71(13-14),1809-1840.

Cheal E. J., Mansmann K. A., DiGioia III A. M., Hayes W. C. and Perren S. M. (1991) Role of interfragmentary strain in fracture healing: ovine model of a healing osteotomy. *J. Orthop. Res.* 9:1, 131-142.

Claes, L. E., Heigele, C. A., (1999) Magnitudes of local stress and strain along bony surfaces predict the course and type of fracture healing.J. Biomech.32:3, 255-265.

Cowin S.C. (1999) Bone poroelasticity. J Biomech, 32:217-238.

Davy, D. T. and Connolly, J. F. (1982) The biomechanical behaviour of healing canine radii and ribs. *J. of Biomech.* 15:4, 235-247.

DiGioia III, A. M., Cheal, E. J., and Hayes, W. C. (1986) Three-dimensional strain fields in a uniform osteotomy gap. *J. Biomech. Eng.* 108, 273-280.

Gardner T. N, Stoll, T, Marks, L, Knothe-Tate, M. (1998) Mathematical modelling of stress and strain in bone fracture repair tissue. *Computer Methods in Biomechanics and Biomedical Engineering* Ed. by J. Middleton and GN Pandy, *Gordon and Breach Science Publishers.* 2, 247 - 254.

Gardner T.N., Marks L., Stoll T., Mishra S., Knothe Tate M., Simpson H.(2000), The influence of mechanical stimulus on the pattern of tissue differentiation in a long bone fracture - An FEM study. J. Biomechanics.33:415-25.

Kenwright, J., and Goodship, A. E. (1989) Controlled mechanical stimulation in the treatment of tibial fractures. *Clin. Orth. Rel. Res.* 241, 36-47.

Kenwright, J. and Gardner, T.N. (1998). Mechanical influences on tibial fracture healing. *Clinical Orthopaedics and Related Research.* 355S, 179-190.

Markel, M. D, Wilkenheiser, M. A. and Chao, E. Y. S. (1990) A study of fracture callus material properties: Relationship to the torsional strength of bone. *J. Orthop. Res.*, 8:6, 843-850.

Mow VC, Kuei SC, Lai WM, Armstrong C. Biphasic creep and stress relaxation of articular cartilage in compression: theory and experiments. J Biomech Engg;1980. 102:73-84.

Oh Jong-Keon, Sahu D, Yoon-Ho Ahn, et al. (2010). Effect of fracture gap on stability of compression plate fixation: A finite element study. J Orthop. Res. 28 (4): 462–467.

Prendergast, P. J., Huiskes, R. and Soballe, K. (1997) Biophysical stimuli on cells during tissue differentiation at implant interfaces. *J. of Biomech.* 30:6, 539–548.

Sarmiento, A. and Latta, l. l. (1995) Functional Fracture Bracing. Tibia, Humerus and Ulna. Springer-Verlag, Berlin.

Simon BR, Wu JSS, Carlton MW, Evans JH, Kazarian LE. Structural models for human spinal motion segments based on poroelastic view of the intervertebral disc. J Biomech Engg; 1985. 107:327-335.

Simon B. Multiphase poroelastic finite element models for soft tissue structures. Appl Mech Rev; 1992: 45(6):191-218

Simon, B. R. (1990) Poroelastic finite element models for soft tissue structures. In *Connective Tissue Matrix, Part 2* (Edited by Hukins, D.), pp. 66-90. MacMillan Press Ltd., London.

Spilker, R. L., Suh, J. K., Vermilyea, M E. and Maxian, T. A. (1990) Alternate hybrid, mixed and penalty finite formulations for the biphasic model of soft hydrated tissues. In *Biomechanics of Diarthrodial Joints* (Edited by Ratcliffe, A. and Woo, S. L-Y.), pp. 400-435. Springer, New York.

Van Driel WD, Huiskes R, Prendergast PJ. A regulatory model for tissue differentiation using poroelastic theory. In 'Poromechanics'; Ed by Thimus JF, Abousleiman Y, Cheng AHD, Coussy O, Detournay E; Publisher A A Balkema Rotterdam, The Netherlands, 1998; 409-413.

Wood D.M. (1990) Soil behaviour and critical state soil mechanics. Cambridge University Press, Cambridge (UK).

Yamagishi, M., and Yoshimura, Y. (1955) The biomechanics of fracture healing. *J. Bone Joint. Surg.* 37A, 1035-1068.

Zienkiewicz, O. C. and Bettess, P. (1982) Soils and other saturated media under transient, dynamic conditions; general formulations and the validity of various simplifying assumptions. In Soil Mechanics - *Transient and Cyclic Loads* (Edited by Pandy, G. N. and Zienkiewicz, O. C.) 1-15. John Wiley and Sons Ltd, UK.

Zienkiewicz O.C., Taylor R.L. (1994a) The Finite Element Method, Vol 1, Mcgraw Hill book Co.

Zienkiewicz O.C., Taylor R.L. (1994b) The Finite Element Method, Vol 2, Mcgraw Hill book Co.

5

Dynamically Incompressible Flow

Christopher Depcik and Sudarshan Loya
University of Kansas, Department of Mechanical Engineering
USA

1. Introduction

Quite often, researchers model a flow as dynamically incompressible without realizing it. This version of the governing equations has been employed to model exhaust aftertreatment devices since the initial work of Vardi and Biller in 1968 (Vardi & Biller, 1968). The small channels in these devices, along with a relatively low flow rate of exhaust gases coming from the engine, promote laminar flow with a speed of approximately one to ten meters per second. This speed is well below the compressible threshold of around 100 m/s or a Mach number of 0.3. As a result, the chemical species equations can be decoupled from the energy equation promoting a computationally faster and easier to program numerical model. While this assumption is indeed valid in this example, only a few researchers have directly stated that the gas is being modeled as dynamically incompressible (Byrne & Norbury, 1993; Depcik et al., 2010). In fact, when this concept is mentioned, reviews of the main author's submitted work in this realm often come back confused as to its meaning. The reviewers wonder how a gas, which is inherently compressible, can be modeled as incompressible. The key wording for this type of situation is "dynamically" incompressible. To clarify this situation, this chapter provides a thorough investigation into this modeling phenomenon.

The efforts begin by explaining the threshold by which a gas can be treated as dynamically incompressible. Dynamic incompressibility differs from incompressibility ($\rho = $ constant) by the fact that the density of the gas is considered as being *approximately* constant ($\rho \approx $ constant). This small change in equal to approximately equal sign has a large bearing on the results. From this assumption, the governing equations of flow are re-derived adding in this assumption in order to provide the proper fundamental versions of these equations for modeling. Of significant importance, when the Law of Conservation of Energy is formulated, an Energy Equation paradox ensues. In particular, two apparently equally valid versions of this equation are found. From this result, this work provides a unique insight into this paradox and indicates the correct description. Moreover, the governing equation of chemical species is included in this chapter, as it not often presented in fluid mechanics books; however, it is important for modeling chemically reactive flow, such as the situation with catalytic exhaust aftertreatment devices.

After describing the governing equations, the failure of this approximation is presented in order for the reader to understand when a model can provide inaccurate results. Finally, this chapter documents a specific example that can lead to highly erroneous results if a modeler does not comprehend the influence of the dynamically incompressible assumption on reacting flows. As a result, this chapter will provide a helpful tool for others when they begin their usage of this version of the fluid mechanics equations.

2. Law of conservation of mass

Conservation of mass is derived from a differential control volume as indicated in Fig. 1. The differential amount of mass (δm) given within the control volume is a function of the product of the fluid density (ρ) and volume (V):

$$\delta m = \rho \delta V \tag{1}$$

with the differential volume indicated as:

$$\delta V = dxdydz \tag{2}$$

It is important to note that all derivations that follow in this chapter, the differential control volume does not change with respect to time.

At each side of this control volume, a mass flux enters and exits the respective control surfaces as a function of local velocity conditions where u, v, and w represent the velocity in the x, y, and z directions respectively. The net mass flux is represented in the x-direction as shown in Fig. 1 as:

$$\text{Net Mass Flux}(x) = \left[\left(\rho u + \frac{\partial(\rho u)}{\partial x} dx \right) - \rho u \right] dydz = \frac{\partial(\rho u)}{\partial x} dxdydz \tag{3}$$

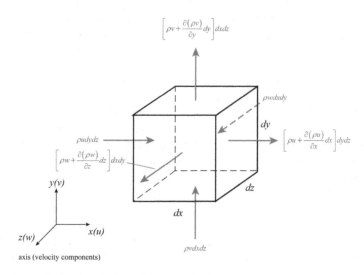

axis (velocity components)

Fig. 1. Mass fluxes through a differential control volume.

In other words, the mass flux changes over the differential control surface in each direction are written as a function of its respective derivative in that direction. Hence, the change in mass flux in the x-direction in the previous equation is equal to:

$$\Delta(\rho u) = \frac{\partial(\rho u)}{\partial x} dx \tag{4}$$

with partial derivatives indicated because each variable is a function of all three dimensions. The change in slope of the x-direction mass flux is multiplied over the distance which the slope changes in order to calculate the difference in mass flux in that respective direction.

In the absence of nuclear reactions, conservation of mass states that there will not be a change in mass encompassing the control volume. Hence, the difference of mass within the control volume as a function of time is balanced by the net flow of mass through the control volume during a certain amount of time:

$$\frac{\delta m}{\delta t} + \sum \text{Net Mass Flux} = 0 \tag{5}$$

Written in differential format, this equation becomes:

$$\frac{\partial \rho}{\partial t} \delta \mathbb{V} + \left[\frac{\partial(\rho u)}{\partial x} dx \right] dy dz + \left[\frac{\partial(\rho v)}{\partial y} dy \right] dx dz + \left[\frac{\partial(\rho w)}{\partial z} dz \right] dx dy = 0 \tag{6}$$

Dividing through by the differential control volume, the conservation of mass is described as:

$$\frac{\partial \rho}{\partial t} + \frac{\partial(\rho u)}{\partial x} + \frac{\partial(\rho v)}{\partial y} + \frac{\partial(\rho w)}{\partial z} = 0 \tag{7}$$

This can be written in vector format for simplicity as:

$$\frac{\partial \rho}{\partial t} + \nabla \cdot (\rho \mathbf{V}) = 0 \tag{8}$$

where the vector $\mathbf{V}$ is equal to $[u \ v \ w]$;

This version of the conservation of mass is known as the conservative formulation as all variables are embedded within the partial derivatives. However, it is often represented in non-conservative format by first expanding the derivatives:

$$\frac{\partial \rho}{\partial t} + \rho \frac{\partial u}{\partial x} + u \frac{\partial \rho}{\partial x} + \rho \frac{\partial v}{\partial y} + v \frac{\partial \rho}{\partial y} + \rho \frac{\partial w}{\partial z} + w \frac{\partial \rho}{\partial z} = 0 \tag{9}$$

and incorporating the definition of divergence in velocity along with the material derivative:

$$\frac{D\rho}{Dt} + \rho \nabla \cdot \mathbf{V} = 0 \tag{10}$$

Both versions of the conservation of mass have their place in numerical analysis for incompressible and compressible flows. In this next section, the criterion for incompressible flow is described.

3. Mach number criterion

A fluid can either be considered incompressible or compressible depending on the interrelationship between pressure, density, velocity and temperature. Liquids and solids are nearly always incompressible as their density remains relatively constant independent of variations in these variables. An exact incompressible fluid has the properties of an infinite sound speed, where a pressure disturbance is felt everywhere within the fluid at each instant in time. Gases are compressible since their density changes as a function of pressure and temperature; often related through the ideal gas law. However, a fluid can be considered *dynamically incompressible* when the speed of sound (aka low-amplitude pressure waves) is significantly faster than the velocity of the working fluid. In this case, the pressure and temperature are not directly related to the density of the medium and a simplification of the governing equations can occur (Kee et al., 2003).

In the study of fluid dynamics, aerodynamicists define a non-dimensional parameter after Ernst Mach that relates the fluid velocity to the speed of sound:

$$M = \frac{V}{a} \tag{11}$$

This parameter helps normalize different working fluids and relate their effects to different flow conditions. As a result, here it will help designate the conditions of incompressible flow independent of the working fluid.

The conservation of mass, equation (8), assuming incompressible flow, results in the following governing equation:

$$\nabla \cdot (\rho V) = 0 \tag{12}$$

By explaining in one-dimension for simplicity, the advection component now equals:

$$\nabla \cdot (\rho V) \rightarrow \frac{\partial(\rho u)}{\partial x} = 0 \tag{13}$$

and further expanding the derivative in non-conservative format recovers:

$$\frac{\partial(\rho u)}{\partial x} = \rho \frac{\partial u}{\partial x} + u \frac{\partial \rho}{\partial x} = 0 \tag{14}$$

For density to be constant, the second term must be small compared to the first term:

$$u \frac{\partial \rho}{\partial x} \ll \rho \frac{\partial u}{\partial x} \Rightarrow \frac{\delta \rho}{\rho} \ll \frac{\delta u}{u} \tag{15}$$

A process of similarity can be used to compare both differential components in equation (15), given small changes in their respective components. This equation indicates that compressibility needs to be considered only when velocity variations are responsible for density variations.

The speed of sound (a) is given by the following thermodynamic relationship under isentropic situations:

$$a^2 = \left(\frac{\partial p}{\partial \rho}\right)_s \approx \frac{\delta p}{\delta \rho} \tag{16}$$

which is then converted similarly to equation (15) into small changes of the respective variables.

Pressure is then related to the velocity by using Bernoulli's equation for irrotational, one-dimensional flow when elevation changes are negligible:

$$\delta p \approx -\rho u \delta u \tag{17}$$

Combining the three equations elucidates the condition for velocity below which flow can be considered to be incompressible:

$$\frac{u^2}{a^2} = M^2 \ll 1 \tag{18}$$

Experimental observations indicate that the following condition can typically be assumed to indicate the limit of dynamically incompressible flow:

$$M \le 0.3 \tag{19}$$

Therefore, when the Mach number is relatively small, it is reasonable to utilize the dynamic incompressibility assumption for the governing equations of fluid dynamics. From this result, the governing equation of continuity is simplified from equation (10) to:

$$\nabla \cdot V = 0 \tag{20}$$

The divergence of velocity in this equation $(\nabla \cdot V)$ is the time rate change of volume of a moving fluid element, per unit volume. Equation (20) means that the change in volume (volume dilation) of the fluid element is zero. This result is consistent with equation (1); i.e., if density and mass are constant, the volume is required to be constant. Written out explicitly using directional variables, the divergence of velocity is given as:

$$\nabla \cdot V = \frac{\partial u}{\partial x} + \frac{\partial v}{\partial y} + \frac{\partial w}{\partial z} = 0 \tag{21}$$

4. Law of conservation of momentum

Consider a fluid element subjected to body forces caused by gravitation, and to surface forces like pressure and shear stresses caused by fluid friction as shown in Fig. 2. From Newton's second law of motion, the time rate of change of momentum of a body (mV) equals the net force (F) exerted on it:

$$F = \frac{d}{dt}(mV) \tag{22}$$

where the left hand side represents the forces exerted on the body and the right hand side is the time rate of change and net flow of momentum within the control volume.

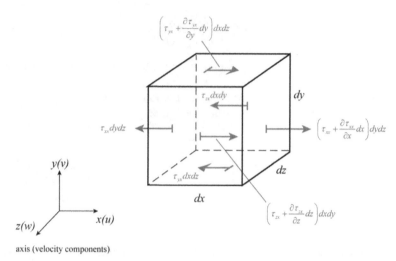

axis (velocity components)

Fig. 2. Forces acting on a control volume, with only x-direction illustrated.

In the absence of external forces, the time rate of change of momentum inside the control volume would be balanced by the momentum flux entering the control volume during a certain amount of time (analogous to the conservation of mass):

$$\frac{\delta(m\mathbf{V})}{\delta t} + \sum \text{Net Momentum Flux} = 0 \qquad (23)$$

Since the mass entering the control volume on each side brings along with it a respective amount of momentum from all three directions, the momentum flux can be derived in a similar manner as the net mass flux as:

$$\text{Net Momentum Flux}(x) = \left[\left(\rho u\mathbf{V} + \frac{\partial(\rho u\mathbf{V})}{\partial x}dx\right) - \rho u\mathbf{V}\right]dydz = \frac{\partial(\rho u\mathbf{V})}{\partial x}dxdydz \qquad (24)$$

For example, the x-momentum equation would equal, in the absence of external forces:

$$\frac{\partial(\rho u)}{\partial t}\delta\mathbb{V} + \left[\frac{\partial(\rho uu)}{\partial x}dx\right]dydz + \left[\frac{\partial(\rho uv)}{\partial y}dy\right]dxdz + \left[\frac{\partial(\rho uw)}{\partial z}dz\right]dxdy = 0 \qquad (25)$$

utilizing equations (1) and (23).

Hence, the total change of momentum inside the control volume for all three-dimensions via the right hand side of equation (22) equals:

$$\frac{d}{dt}(m\mathbf{V}) \rightarrow \frac{\partial(\rho\mathbf{V})}{\partial t}\delta\mathbb{V} + (\nabla \cdot \rho\mathbf{VV})\delta\mathbf{S} \qquad (26)$$

where $\delta\mathbf{S}$ is a vector indicating the control surface in the respective direction:

$$\delta S_x = dydz \; ; \; \delta S_y = dxdz \; ; \; \delta S_z = dxdy \qquad (27)$$

The forces on this fluid element are broken up into those that act at a distance, caused by force fields acting through space proportional to the control volume (body), and shear and strain forces (surface) that include both normal and tangential components proportional to the area:

$$\mathbf{F} = \mathbf{F}_B + \mathbf{F}_S \tag{28}$$

where $\mathbf{F}_B$ are body forces and $\mathbf{F}_S$ are surface forces in the above equation.
Body forces in this case include only gravity:

$$\mathbf{F}_B = (\rho \delta \mathbb{V})\mathbf{g} \tag{29}$$

where $\mathbf{g}$ is a vector oriented in the direction of the solved equation. Other body forces do exist like electromagnetic forces; however, these forces are beyond the scope of this chapter.
The surface forces act on the boundary of the control volume and are applied by external stresses on the sides of the element. Similar to the net momentum flux, a net stress component can be derived as indicated via Fig. 2 and equation (24)

$$\text{Net Stress}(x) = \frac{\partial \tau_{xx}}{\partial x} dxdydz + \frac{\partial \tau_{yx}}{\partial y} dydxdz + \frac{\partial \tau_{zx}}{\partial z} dzdxdy \tag{30}$$

where the first subscript on the stress component (τ) indicates the surface orientation by providing the direction of its outward normal and the second subscript indicates the direction of the force component.
In the absence of shear and strain, flow can and will happen because of the normal compressive stress exerted by the working fluid on the boundaries. This is called the hydrostatic pressure condition, as pressure force always acts normal and against the control surface in each direction. Hence, the stress tensor can be represented as:

$$\tau = -p\delta_{ij} + \begin{bmatrix} \tau_{xx} & \tau_{xy} & \tau_{xz} \\ \tau_{yx} & \tau_{yy} & \tau_{yz} \\ \tau_{zx} & \tau_{zy} & \tau_{zz} \end{bmatrix} = -p\delta_{ij} + \tau' \tag{31}$$

with the complete surface force represented as:

$$\delta \mathbf{F}_S = \tau \cdot \delta \mathbf{S} \tag{32}$$

The rest of the stress tensor in equation (31) is calculated by considering the motion of a fluid element as it moves through a flow field. Fluids can undergo a number of different phenomena, as indicated in Fig. 3:

- Translation – linear movement from one location to another
- Rotation – the sides of the element may change as a function of pure rotation
- Angular deformation – distortion of the element by converting perpendicular planes to non-perpendicular planes
- Linear deformation – a change in shape without a change in orientation
- Volume dilatation – rate of change of volume per unit volume

Of these different influences, the last three relate to the strain that the fluid element undergoes. Only a brief summary of these components is presented in this section and the readers may wish to refer to the following references for a full derivation (Anderson, 1995; Schlichting & Gersten, 2000; Fox et al., 2004; White, 2003; Panton, 2005; White, 2006).

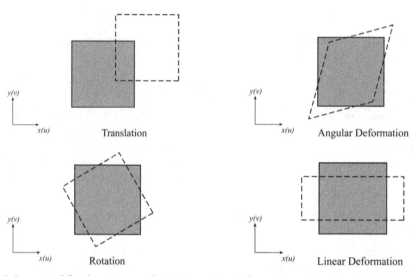

Fig. 3. Impact of fluid motion on fluid element as it flows through a flow field in two-dimensions.

Angular deformation occurs when the fluid element experiences a rate of deformation given by the change in velocity over the distance computed. In order to formulate a mathematical interpretation of this phenomenon, two assumptions are often made. The first relates to the assumption of a Newtonian Fluid, where the stress at a point is linearly dependent on the rates of strain (deformation) of the fluid. The second involves classifying the fluid as isotropic where the properties of the fluid are independent of direction of orientation. As a result, the components in the stress tensor that are a function of angular deformation are:

$$\tau_{yz} = \tau_{zy} = \mu\left(\frac{\partial w}{\partial y} + \frac{\partial v}{\partial z}\right); \ \tau_{xy} = \tau_{yx} = \mu\left(\frac{\partial u}{\partial y} + \frac{\partial v}{\partial x}\right); \ \tau_{zx} = \tau_{xz} = \mu\left(\frac{\partial u}{\partial z} + \frac{\partial w}{\partial x}\right) \quad (33)$$

where μ is a constant of proportionality that can be measured experimentally; also known as the dynamic viscosity.

Because of velocity gradients, the fluid element can deform as it moves. Linear deformation happens when the flow causes strain by stretching or shrinking the shape of the fluid element. Because of the isotropic condition previously specified, it was found that the linear coefficient of proportionality is equal to twice the constant of proportionality of angular deformation:

$$\tau_{xx,a} = 2\mu\frac{\partial u}{\partial x}; \ \tau_{yy,a} = 2\mu\frac{\partial v}{\partial y}; \ \tau_{zz,a} = 2\mu\frac{\partial w}{\partial z} \quad (34)$$

The fluid can also undergo a volumetric dilatation that equals the rate of change of the volume per unit volume. This component can be thought as the elasticity or compressibility of the working fluid, and while volume dilation and linear deformation are similar concepts, they are not the same. For example, a fluid element can undergo linear deformation (change in shape) but not necessarily volume dilation (change in volume). Similar to linear deformation, volume dilation does not involve any angular deflection and is purely linear in nature. Of interest, even though the fluid volume might only change in the x-direction, it does impact the other momentum equations. This is because the time-change component of the other momentum equations via equation (26) includes the density of the working fluid. Hence, even though the volume is changing only in the x-direction, it will influence the other governing equations through the density dependence. Moreover, because the isotropic condition is applied, the deformation is independent of the coordinate axis under which it is expressed. Therefore, the strain will impact all dimensions; hence, the normal component will be equal among all three dimensions:

$$\tau_{xx,b} = \tau_{yy,b} = \tau_{zz,b} = \lambda\left(\frac{\partial u}{\partial x} + \frac{\partial v}{\partial y} + \frac{\partial w}{\partial z}\right) = \lambda(\nabla \cdot \mathbf{V}) \tag{35}$$

The coefficient λ was originally considered independent of μ and is often called the second coefficient of viscosity in order to help differentiate volume dilation from angular deformation. In linear elasticity according to a Newtonian description, this variable is called Lamé's constant. For estimation of this variable, consult Gad-el-Hak for an understanding of the assumptions present in its calculation (Gad-el-Hak, 1995).

At this point, all phenomena have been developed and the momentum equations can be written out explicitly. The x-momentum equation expands to:

$$\frac{\partial(\rho u)}{\partial t}\delta\mathbb{V} + \left[\frac{\partial(\rho uu)}{\partial x}dx\right]dydz + \left[\frac{\partial(\rho uv)}{\partial y}dy\right]dxdz + \left[\frac{\partial(\rho uw)}{\partial z}dz\right]dxdy =$$
$$(\rho\delta\mathbb{V})g_x - \left(\frac{\partial p}{\partial x}dx\right)dydz + \left(\frac{\partial\tau_{xx}}{\partial x}dx\right)dydz + \left(\frac{\partial\tau_{yx}}{\partial y}dy\right)dxdz + \left(\frac{\partial\tau_{zx}}{\partial z}dz\right)dydz \tag{36}$$

and dividing through by the volume while making the assumption of incompressibility recovers:

$$\rho\left[\frac{\partial u}{\partial t} + u\frac{\partial u}{\partial x} + v\frac{\partial u}{\partial y} + w\frac{\partial u}{\partial z} + u\underbrace{\left(\frac{\partial u}{\partial x} + \frac{\partial v}{\partial y} + \frac{\partial w}{\partial z}\right)}_{\nabla\cdot\mathbf{V}=0}\right] = \rho g_x - \frac{\partial p}{\partial x} + \frac{\partial\tau_{xx}}{\partial x} + \frac{\partial\tau_{yx}}{\partial y} + \frac{\partial\tau_{zx}}{\partial z} \tag{37}$$

Simplifying and incorporating the stress components equals:

$$\rho\frac{Du}{Dt} = \rho g_x + \frac{\partial}{\partial x}\left[-p + 2\mu\frac{\partial u}{\partial x}\right] + \frac{\partial}{\partial y}\mu\left(\frac{\partial u}{\partial y} + \frac{\partial v}{\partial x}\right) + \frac{\partial}{\partial z}\mu\left(\frac{\partial u}{\partial z} + \frac{\partial w}{\partial x}\right) \tag{38}$$

Note that Lamé's constant does not appear in the above equation because the velocity gradient via equation (20) is approximately equal to zero. Therefore, the influence of Lamé's constant (λ) is ignored as it is multiplied by a negligible term; e.g. viscous stress is much smaller in magnitude than other flow parameters, like pressure, which results in the multiplication of a relatively small term by a negligible term.

For a Newtonian fluid, viscosity depends on temperature and pressure. In the case of a dynamically incompressible fluid, as discussed later, the change of temperature and pressure across the region must be relatively small in order for the assumption of dynamic incompressibility to hold. Hence, there is a negligible change in viscosity as a function of these parameters and the assumption of constant viscosity is valid.

$$\rho\frac{Du}{Dt} = -\frac{\partial p}{\partial x} + \mu\underbrace{\left[\frac{\partial^2 u}{\partial x^2} + \frac{\partial^2 u}{\partial y^2} + \frac{\partial^2 u}{\partial z^2}\right]}_{\nabla^2 u} + \mu\left[\frac{\partial^2 u}{\partial x^2} + \frac{\partial^2 v}{\partial y\partial x} + \frac{\partial^2 w}{\partial z\partial x}\right] + \rho g_x \tag{39}$$

By collecting the derivative in the second to last term on the right hand side, this results in:

$$\rho\frac{Du}{Dt} = -\frac{\partial p}{\partial x} + \mu\nabla^2 u + \mu\frac{\partial}{\partial x}\underbrace{\left[\frac{\partial u}{\partial x} + \frac{\partial v}{\partial y} + \frac{\partial w}{\partial z}\right]}_{\nabla\cdot V = 0} + \rho g_x \tag{40}$$

Hence, this term disappears since the derivative of the velocity gradient is approximately equal to zero from dynamic incompressibility. Therefore, the final x-momentum governing equation for a dynamically incompressible flow equals:

$$\rho\frac{Du}{Dt} = -\frac{\partial p}{\partial x} + \mu\nabla^2 u + \rho g_x \tag{41}$$

Similarly, the momentum equations for dynamic incompressible flow in the y and z directions can be obtained as:

$$\rho\frac{Dv}{Dt} = -\frac{\partial p}{\partial y} + \mu\nabla^2 v + \rho g_y \tag{42}$$

$$\rho\frac{Dw}{Dt} = -\frac{\partial p}{\partial z} + \mu\nabla^2 w + \rho g_z \tag{43}$$

5. Law of conservation of energy

The physical principle governing the law of conservation of energy is that the total energy of the system must be conserved. Similar to the last two sections, consider a small fluid element moving with the fluid flow as in Fig. 4. The rate of change of total energy inside the fluid element (E) will be equal to the addition of the net heat flux (Q) into the element and rate of work done (W) on the fluid element due to body and shear forces. In mathematical form, this is represented as:

$$\frac{d}{dt}(mE) = \frac{dQ}{dt} - \frac{dW}{dt} \tag{44}$$

where heat transfer is defined as positive into the control volume and work is defined as positive out of the control volume.

The total energy inside the fluid element is a function of its internal energy, potential and kinetic energy due to the translational motion of the fluid element:

$$E = e + \tfrac{1}{2}\tilde{V}^2 + \mathbf{g} \tag{45}$$

where the effective kinetic energy velocity incorporates all three components of direction:

$$\tilde{V}^2 = u^2 + v^2 + w^2 \tag{46}$$

In the absence of heat transfer and work, the time rate of change of energy inside the control volume would be balanced by the energy flux entering the control volume during a certain amount of time (analogous to the conservation of mass):

$$\frac{\delta(mE)}{\delta t} + \sum \text{Net Energy Flux} = 0 \tag{47}$$

Since the mass entering the control volume on each side brings along with it a respective amount of energy from all three directions, the energy flux can be derived in a similar manner as the net mass flux as:

$$\text{Net Energy Flux}(x) = \left[\left(\rho u E + \frac{\partial(\rho u E)}{\partial x} dx \right) - \rho u E \right] dy dz = \frac{\partial(\rho u E)}{\partial x} dx dy dz \tag{48}$$

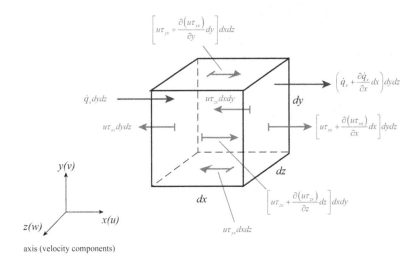

axis (velocity components)

Fig. 4. Energy fluxes through an infinitesimally small fluid element with only the fluxes in the x-direction illustrated.

Hence, written in differential format, the left hand side of equation (44) becomes:

$$\frac{d}{dt}(mE) \rightarrow \frac{\partial(\rho E)}{\partial t}\delta\mathbb{V} + \left[\frac{\partial(\rho u E)}{\partial x}dx\right]dydz + \left[\frac{\partial(\rho v E)}{\partial y}dy\right]dxdz + \left[\frac{\partial(\rho w E)}{\partial z}dz\right]dxdy \qquad (49)$$

The net heat flux, as indicated in Fig. 4, incorporates a negative sign in order to account for the definition that heat transfer into the control volume is positive:

$$\text{Net Heat Flux}(x) = -\left[\left(\dot{q}_x + \frac{\partial \dot{q}_x}{\partial x}dx\right) - \dot{q}_x\right]dydz = -\frac{\partial \dot{q}_x}{\partial x}dxdydz \qquad (50)$$

Based on phenomenological evidence, Fourier found that the heat transfer rate in the direction per unit area perpendicular to the direction of transfer is linearly proportional to the temperature gradient in this direction. Furthermore, through the isotropic condition previously mentioned, all three directions incorporate the same proportionality constant, k, in determination of the heat flux:

$$\dot{q}_x = -k\frac{\partial T}{\partial x}; \; \dot{q}_y = -k\frac{\partial T}{\partial y}; \; \dot{q}_z = -k\frac{\partial T}{\partial z} \qquad (51)$$

This constant, referred hereto as thermal conductivity with a value that is a function of time, is analogous in its concept to viscosity in the momentum equation. Note that if the derivative of temperature is negative, heat transfer is positive as it is propagating in the forward direction (moving from hot to cold). Hence, the total net heat flux is equal to:

$$\frac{dQ}{dt} \rightarrow \left\{\frac{\partial}{\partial x}\left(k\frac{\partial T}{\partial x}\right) + \frac{\partial}{\partial y}\left(k\frac{\partial T}{\partial y}\right) + \frac{\partial}{\partial z}\left(k\frac{\partial T}{\partial z}\right)\right\}dxdydz \qquad (52)$$

The work done on a fluid element is a function of the forces acting on this element. As indicated in the momentum equation, there are two forces (body and surface) evident. The rate of work done by a force is the product of this force and the component of velocity in the direction of the force. In order words, the time rate of work is equal to the force done over a certain distance as a function of time. Hence, the work done by the body force is represented as:

$$\text{Body Work} = (\rho\delta\mathbb{V})\mathbf{g} \cdot \mathbf{V} \qquad (53)$$

The rate of work done by the surface forces are the product of the stresses with the component of velocity in the corresponding direction. For example, the rate of work in the x-direction is equal to:

$$\text{Net Work}(x) = -\frac{\partial(u\tau_{xx})}{\partial x}dxdydz - \frac{\partial(u\tau_{yx})}{\partial y}dydxdz - \frac{\partial(u\tau_{zx})}{\partial z}dzdxdy \qquad (54)$$

including a negative sign as the force on the control volume is opposing the direction of the flow.

Including the work done by pressure and surface forces in all three directions, the total rate of work done is equal to:

$$\frac{dW}{dt} = -\left\{\frac{\partial\left(u\tau_{xx}+v\tau_{xy}+w\tau_{xz}\right)}{\partial x}+\frac{\partial\left(u\tau_{yx}+v\tau_{yy}+w\tau_{yz}\right)}{\partial y}+\frac{\partial\left(u\tau_{zx}+v\tau_{zy}+w\tau_{zz}\right)}{\partial z}\right\}dxdydz \quad (55)$$

And the governing equation for the conservation of energy, incorporating equations (49), (52) and (55) into equation (44) while dividing by the volume, equals:

$$\frac{\partial(\rho E)}{\partial t}+\frac{\partial(\rho u E)}{\partial x}+\frac{\partial(\rho v E)}{\partial y}+\frac{\partial(\rho w E)}{\partial z}=$$

$$\frac{\partial}{\partial x}\left(k\frac{\partial T}{\partial x}\right)+\frac{\partial}{\partial y}\left(k\frac{\partial T}{\partial y}\right)+\frac{\partial}{\partial z}\left(k\frac{\partial T}{\partial z}\right)+ \quad (56)$$

$$\frac{\partial\left(u\tau_{xx}+v\tau_{xy}+w\tau_{xz}\right)}{\partial x}+\frac{\partial\left(u\tau_{yx}+v\tau_{yy}+w\tau_{yz}\right)}{\partial y}+\frac{\partial\left(u\tau_{zx}+v\tau_{zy}+w\tau_{zz}\right)}{\partial z}+\rho(\mathbf{g}\cdot\mathbf{V})$$

which can be represented in vector format as:

$$\frac{\partial(\rho E)}{\partial t}+\nabla\cdot(\rho\mathbf{V}E)=\nabla\cdot(k\nabla T)-\nabla(p\mathbf{V})+\nabla(\mathbf{V}\cdot\tau')+\rho(\mathbf{g}\cdot\mathbf{V}) \quad (57)$$

with the pressure term separated out from the stress tensor as doing so will become important in a later section.

As a result, the energy equation governing fluid flow consists of two energy sources. Work through the body forces accelerates the fluid and increases its kinetic energy, while thermal energy conduction (heat flux) increases its internal energy (Panton, 2005). This equation can be further simplified by expanding the left hand side:

$$\rho\frac{\partial E}{\partial t}+\rho\mathbf{V}\cdot\nabla E+E\underbrace{\left[\frac{\partial\rho}{\partial t}+\nabla\cdot(\rho\mathbf{V})\right]}_{0}=\nabla\cdot(k\nabla T)-\nabla(p\mathbf{V})+\nabla(\mathbf{V}\cdot\tau')+\rho(\mathbf{g}\cdot\mathbf{V}) \quad (58)$$

and incorporating the conservation of mass via equation (8) in order to recover:

$$\rho\frac{DE}{Dt}=\nabla\cdot(k\nabla T)-\nabla(p\mathbf{V})+\nabla(\mathbf{V}\cdot\tau')+\rho(\mathbf{g}\cdot\mathbf{V}) \quad (59)$$

Of importance, when the mechanical work component is subtracted from the total energy equation, the remaining part is called the thermal energy equation. As discussed earlier, mechanical work is equal to product of force and velocity. All the forces acting on the body are described via the momentum equation. Hence, in order to obtain the mechanical energy equation in the x-direction, one can multiply the momentum equation (37) (before simplifying this equation further) by the respective velocity component u as follows:

$$\rho\frac{D\left(u^2/2\right)}{Dt}=\left[-u\frac{\partial p}{\partial x}+u\frac{\partial\tau_{xx}}{\partial x}+u\frac{\partial\tau_{yx}}{\partial y}+u\frac{\partial\tau_{zx}}{\partial z}+\rho u g_x\right] \quad (60)$$

Similarly, the mechanical energy equations for the y and z directions can be found as:

$$\rho\frac{D\left(v^2/2\right)}{Dt} = \left[-v\frac{\partial p}{\partial y} + v\frac{\partial \tau_{xy}}{\partial x} + v\frac{\partial \tau_{yy}}{\partial y} + v\frac{\partial \tau_{zy}}{\partial z} + \rho v g_y \right] \tag{61}$$

$$\rho\frac{D\left(w^2/2\right)}{Dt} = \left[-w\frac{\partial p}{\partial z} + w\frac{\partial \tau_{xz}}{\partial x} + w\frac{\partial \tau_{yz}}{\partial y} + w\frac{\partial \tau_{zz}}{\partial z} + \rho w g_z \right] \tag{62}$$

By adding these equations, the total mechanical energy is obtained:

$$
\begin{aligned}
\rho\frac{D\left(\dfrac{\tilde{V}^2}{2}\right)}{Dt} = & -\left(u\frac{\partial p}{\partial x} + v\frac{\partial p}{\partial y} + w\frac{\partial p}{\partial z} \right) + u\left(\frac{\partial \tau_{xx}}{\partial x} + \frac{\partial \tau_{yx}}{\partial y} + \frac{\partial \tau_{zx}}{\partial z} \right) \\
& + v\left(\frac{\partial \tau_{xy}}{\partial x} + \frac{\partial \tau_{yy}}{\partial y} + \frac{\partial \tau_{zy}}{\partial z} \right) + w\left(\frac{\partial \tau_{xz}}{\partial x} + \frac{\partial \tau_{yz}}{\partial y} + \frac{\partial \tau_{zz}}{\partial z} \right) \\
& + \underbrace{\rho u g_x + \rho v g_y + \rho w g_z}_{\rho(\mathbf{g}\cdot\mathbf{V})}
\end{aligned}
\tag{63}
$$

which in vector format equals:

$$\rho\frac{D}{Dt}\left(\frac{\tilde{V}^2}{2}\right) = -\mathbf{V}\cdot\nabla p + \mathbf{V}\cdot\nabla\cdot\boldsymbol{\tau}' + \rho(\mathbf{g}\cdot\mathbf{V}) \tag{64}$$

When equation (45) is used in equation (59) and the derivitives are expanded on the right-hand side, the result is:

$$\rho\frac{D(e+\mathbf{g})}{Dt} + \rho\frac{D}{Dt}\left(\frac{\tilde{V}^2}{2}\right) = \nabla\cdot\left(k\nabla T\right) - p\nabla\cdot\mathbf{V} - \mathbf{V}\cdot\nabla p + \mathbf{V}\cdot\nabla\cdot\boldsymbol{\tau}' + \boldsymbol{\tau}':\nabla\mathbf{V} + \rho(\mathbf{g}\cdot\mathbf{V}) \tag{65}$$

The total mechanical energy from equation (64) can now be subtracted from the total energy equation, equation (65), in order to recover the thermal energy equation. Assuming that gravity has a negligible influence, the result is:

$$\rho\frac{De}{Dt} = \nabla\cdot\left(k\nabla T\right) - p\nabla\cdot\mathbf{V} + \boldsymbol{\tau}':\nabla\mathbf{V} \tag{66}$$

In order to utilize the thermal energy equation for modeling purposes, it is customary to convert it into temperature as the dependant variable (White, 2003). This conversion can be accomplished using either of two different methods as illustrated in the next section. As a result, an energy equation paradox results.

6. Energy equation paradox

Internal energy is a thermodynamic property that can be expressed by two fundamental properties of state. Here it is expressed in terms of temperature and specific volume:

$$de = \underbrace{\left(\frac{\partial e}{\partial T}\right)_v}_{c_v} dT + \left(\frac{\partial e}{\partial v}\right)_T dv \tag{67}$$

Note that the first term on the right hand side is the definition of the constant volume specific heat.

Substituting equation (67) in equation (66) finds:

$$\rho c_v \frac{DT}{Dt} + \rho \left(\frac{\partial e}{\partial v}\right)_T \frac{Dv}{Dt} = \nabla \cdot (k\nabla T) - p\nabla \cdot \mathbf{V} + \tau' : \nabla \mathbf{V} \tag{68}$$

However, the specific volume is reciprocal of density and, therefore equation (68) becomes:

$$\rho c_v \frac{DT}{Dt} + \left(\frac{\partial e}{\partial v}\right)_T \left(-\frac{1}{\rho}\right) \frac{D\rho}{Dt} = \nabla \cdot (k\nabla T) - p\nabla \cdot \mathbf{V} + \tau' : \nabla \mathbf{V} \tag{69}$$

Alternatively, internal energy can be expressed in terms of enthalpy as:

$$e = h - p/\rho \tag{70}$$

Written in derivative format, equation (70) becomes:

$$de = dh - dp/\rho + p\left(d\rho/\rho^2\right) \tag{71}$$

Since enthalpy is a thermodynamic property, it can additionally be expressed by two fundamental state variables; here, it is expressed as a function of temperature and pressure:

$$dh = \underbrace{\left(\frac{\partial h}{\partial T}\right)_p}_{c_p} dT + \left(\frac{\partial h}{\partial p}\right)_T dp \tag{72}$$

where the first term on the right hand side is the definition of the constant pressure specific heat of a fluid.

Using the basic law of thermodynamics along with Maxwell's relation,

$$\left(\frac{\partial h}{\partial p}\right)_T = \left[v + T\left(\frac{\partial s}{\partial p}\right)_T\right] = \left[v + T\left(\frac{\partial v}{\partial T}\right)_p\right] = \frac{1}{\rho}\left[1 - \frac{T}{\rho}\left(\frac{\partial \rho}{\partial T}\right)_p\right] \tag{73}$$

And substituting equations (72) and (73) into equation (71), one can write the internal energy as a function of the constant pressure specific heat:

$$de = c_p dT + \frac{1}{\rho}\left[1 - \underbrace{\frac{T}{\rho}\left(\frac{\partial \rho}{\partial T}\right)_p}_{\beta}\right] dp - \frac{dp}{\rho} + \frac{p}{\rho^2} d\rho \tag{74}$$

Note that the derivative of density with respect to temperature is the definition of the thermal expansion coefficient (β). In particular, a substance will expand with an increase in energy (heating) and contract with a decrease (cooling) with this dimensional response expressed as the coefficient of thermal expansion.

Furthermore, incorporating (74) into equation (66):

$$\rho\left[c_p\frac{DT}{Dt}-\frac{T\beta}{\rho}\frac{Dp}{Dt}+\frac{p}{\rho^2}\frac{D\rho}{Dt}\right]=\nabla\cdot(k\nabla T)-p\nabla\cdot\mathbf{V}+\boldsymbol{\tau}':\nabla\mathbf{V}\tag{75}$$

The terms in equation (75) can be rearranged and regrouped to produce:

$$\rho c_p\frac{DT}{Dt}=T\beta\frac{Dp}{Dt}-p\underbrace{\left[\frac{1}{\rho}\frac{D\rho}{Dt}+(\nabla\cdot\mathbf{V})\right]}_{0}+\boldsymbol{\tau}':\nabla\mathbf{V}+\nabla\cdot(k\nabla T)\tag{76}$$

Note that the expression in square brackets on the right hand side is the continuity equation for dynamically incompressible flow in non-conservative format and must therefore be equal to zero. As a result, the final thermal equation following this procedure equals:

$$\rho c_p\frac{DT}{Dt}=T\beta\frac{Dp}{Dt}+\boldsymbol{\tau}':\nabla\mathbf{V}+\nabla\cdot(k\nabla T)\tag{77}$$

As a result, equation (69) and equation (77) both represent the thermal energy equation utilizing temperature as the dependant variable. If one applies the assumption of dynamic incompressibility to both equations, the results are:

$$\rho c_v\frac{DT}{Dt}=\nabla\cdot(k\nabla T)\tag{78}$$

$$\rho c_p\frac{DT}{Dt}=\nabla\cdot(k\nabla T)\tag{79}$$

as the terms that were eliminated are negligible for an incompressible fluid.

In case of truly incompressible liquids and solids, the difference between the specific heat at constant volume and constant pressure vanishes. Hence, equation (78) and (79) reduce to a single equation:

$$\rho c\frac{DT}{Dt}=\nabla\cdot(k\nabla T)\tag{80}$$

However, because the specific heats of gases have distinct values equations (78) and (79) lead to a unique problem called the Energy Equation Paradox (White, 2003; Panton, 2005). Both equation (78) and equation (79) cannot be simultaneously valid for the dynamically incompressible flow of gases. In particular, equation (78) implies that advection of internal energy is balanced by heat conduction, while equation (79) implies that enthalpy advection is balanced by heat conduction. In order to explain this paradox and obtain a solution, one must first understand the assumption of dynamic incompressibility, its conditions and application in the governing equations.

Any gaseous flow is assumed to be incompressible if the velocity is less than a Mach number of 0.3 and there is not a large local change in temperature and pressure. For such a flow condition, density is assumed to be constant $(\rho = \text{constant})$ and the divergence of velocity is set equal to zero $(\nabla \cdot \mathbf{V} = 0)$. However, in actuality, there is a negligible change in density $(\rho \approx \text{constant})$ and divergence of velocity is not quite zero $(\nabla \cdot \mathbf{V} \approx 0)$. As a result, when the temperature gradient is not large, the conduction and advection of gases are relatively small and nearly of same magnitude as the divergence of velocity. Hence, in the thermal energy equation, any term containing the divergence of velocity or substantial derivative of pressure, temperature or density cannot be set immediately to zero without a thorough review.

6.1 Viscous dissipation

Viscous dissipation is always positive and acts to create internal energy (Panton, 2005). This change in energy is irreversible and it is written as a dyadic product of two tensors, shear stress and gradient of velocity $(\tau' : \nabla \mathbf{V})$ resulting in scalar work (Kee et al., 2003). Viscous dissipation describes rate of work for shape change at constant volume. For dynamically incompressible flow, change in shape at constant volume is negligible as density is assumed to be constant; hence, viscous dissipation is relatively small (Kee et al., 2003). Moreover, viscous dissipation becomes important when the fluid is highly viscous or turbulent (Kreith et al., 2010). In this case, the fluid is a gas with low viscosity under low Mach number situations. Therefore, the change in internal energy due to viscous dissipation will not influence the internal energy significantly and, subsequently, the temperature. As a result, it can be neglected in both equations (69) and (77).

6.2 Substantial derivative of pressure

The substantial derivative of pressure factor $\left[T\beta(Dp/Dt) \right]$ includes the thermal expansion coefficient, β. For an ideal gas, this coefficient is the reciprocal of temperature; hence, it cancels out the temperature component leaving just the material pressure derivative (Kundu & Cohen, 2010). For incompressible flow, thermodynamics properties (like k, c_p, μ) are often considered constant. Although they fundamentally change with temperature, one of the overriding assumptions for dynamic incompressibility is that there is not a substantial temperature change. If this assumption does not hold,the flow must be treated as compressible (Panton, 2005; Depcik et al., 2010). This assumption allows decoupling of the continuity and momentum equations from the thermal energy equation. Hence, all three velocities and pressure can be solved without needing to compute the temperature simultaneously. Therefore, the velocity field and pressure are unaffected by thermal changes in incompressible flow, since they are derived from the mass and momentum equations. Thus, pressure is represented as a force and not as a property influencing temperature. If the pressure increases or decreases across the incompressible flow region, the level of all pressures increases or decreases respectively. As a result, the change in pressure across the flow is negligible and it can be eliminated from equation (60).

6.3 Substantial derivative of density

For dynamically incompressible flow, the change in density is negligible. Moreover, any change in internal energy corresponding to a change in volume is marginal because the change in volume itself is small, as indicated in equation (21). Since this component is a product of two trivial terms, it is neglected in equation (69).

6.4 Pressure times divergence of velocity

Although the divergence of velocity may be small, pressure across the flow is significant in magnitude (Kee et al., 2003). Hence, this term is not inconsequential as the product is on the same order of scale as that of conduction: $p(\nabla \cdot \mathbf{V})$. As a result, applying the previous discussions and the influence of this component, the final thermal energy equation is obtained as a function of constant volume or constant pressure specific heats:

$$\rho c_v \frac{DT}{Dt} = \nabla \cdot (k\nabla T) - p(\nabla \cdot \mathbf{V}) \tag{81}$$

$$\rho c_p \frac{DT}{Dt} = \nabla \cdot (k\nabla T) \tag{82}$$

In order to validate the previous discussion, the above two equations must generate the same results. Hence, the following derivation proves the methodology and explains the paradox.

For dynamic incompressibility, the change in density of a particle is negligible and therefore density is assumed constant. In mathematical form, this is represented as:

$$\frac{D\rho}{Dt} = 0 = \frac{1}{\rho}\frac{D\rho}{Dt} = \frac{1}{\rho}\frac{\partial \rho}{\partial p}\bigg|_T \frac{Dp}{Dt} + \frac{1}{\rho}\frac{\partial \rho}{\partial T}\bigg|_p \frac{DT}{Dt} \tag{83}$$

Since density is governed by the thermodynamic equation of state, it can be represented as a function of temperature and pressure (Panton, 2005). In addition, equation (83) can be written as:

$$\frac{1}{\rho}\frac{D\rho}{Dt} = \alpha \frac{Dp}{Dt} - \beta \frac{DT}{Dt} = 0 \tag{84}$$

where the isothermal compressibility 'a' and thermal expansion coefficient, β, are the characteristic of fluid and hence cannot be set to zero (Turcotte & Schubert, 2002; Graetzel & Infelta, 2002). However, if the change in pressure and change in temperature are sufficiently small, then equation (84) is equal to zero.

Incorporating equation (84) into equation (21):

$$-(\nabla \cdot \mathbf{V}) = \alpha\left(\frac{Dp}{Dt}\right) - \beta\left(\frac{DT}{Dt}\right) \tag{85}$$

As a result, equation (81) is modified to:

$$\rho c_v \frac{DT}{Dt} = \nabla \cdot (k\nabla T) + p\left[\alpha\left(\frac{Dp}{Dt}\right) - \beta\left(\frac{DT}{Dt}\right)\right] \tag{86}$$

For an ideal gas, isothermal compressibility 'a' is the reciprocal of pressure and the thermal expansion coefficient, β, is the reciprocal of temperature (Roy, 2001; Honerkamp, 2002; Kundu & Cohen, 2010). Incorporating these simplifications into equation (86) results in:

$$\rho\left(c_v + \frac{\beta}{\rho\alpha}\right)\frac{DT}{Dt} = \nabla \cdot (k\nabla T) + \frac{Dp}{Dt} \tag{87}$$

As explained earlier, the change in pressure across the flow is negligible and the substantial derivative of pressure in this equation is set to zero. Moreover, the specific heats of gases are related by the gas constant as:

$$c_p = c_v + R \Rightarrow c_v + \frac{p}{\rho T} \tag{88}$$

Using the property of isothermal compressibility and thermal expansion coefficient for ideal gases, equation (88) converts to:

$$c_p = c_v + \frac{p}{\rho T} = c_v + \frac{\beta}{\rho\alpha} \tag{89}$$

When equation (89) is used in equation (87), the result is identical to equation (79):

$$\rho c_p \frac{DT}{Dt} = \nabla \cdot (k\nabla T) \tag{79}$$

This alternative development of equation (79) resolves the energy equation paradox; for dynamically incompressible flow, the advection of enthalpy is balanced by conduction.

7. Law of conservation of species

The law of conservation of species follows the same principles as the law of conservation of mass. At each side of the control volume indicated in Fig. 5, a species flux enters and exits the respective control surfaces as a function of local velocity conditions where u_A, v_A, and w_A represent the total species A velocity in the x, y, and z directions, respectively. The net species flux is represented in the x-direction as shown in Fig. 5 as:

$$\text{Net Species Flux}(x) = \left[\left(\rho_A u_A + \frac{\partial(\rho_A u_A)}{\partial x}dx\right) - \rho_A u_A\right]dydz = \frac{\partial(\rho_A u_A)}{\partial x}dxdydz \tag{90}$$

where ρ_A is the density of the individual species considered.

Because reactions of chemical species may occur, the difference of mass of the species within the control volume as a function of time is balanced by the net flow of species through the control volume during a certain amount of time while including a local production or destruction rate:

$$\frac{\delta m_A}{\delta t} + \sum \text{Net Species Flux} = \dot{\omega}_A \delta \mathbb{V} \tag{91}$$

where the differential mass is written as:

$$\delta m_A = \rho_A \delta \mathbb{V} \tag{92}$$

and the chemical reaction rate has the units of kg m^{-3} s^{-1}, consumes species A when negative, and acts over the entire control volume. In differential format, equation (91) becomes:

$$\frac{\partial \rho_A}{\partial t}\delta\mathbb{V} + \left[\frac{\partial(\rho_A u_A)}{\partial x}dx\right]dydz + \left[\frac{\partial(\rho_A v_A)}{\partial y}dy\right]dxdz + \left[\frac{\partial(\rho_A w_A)}{\partial z}dz\right]dxdy = \dot{\omega}_A dxdydz \tag{93}$$

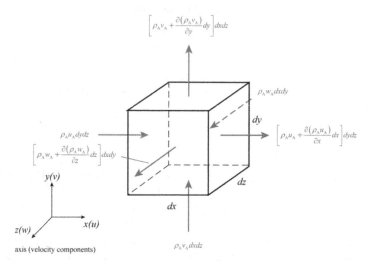

Fig. 5. Species fluxes through a differential control volume.

When equation (93) is divided through by $\delta\mathbb{V} = \delta x \delta y \delta z$, the differential control volume, the conservation of species A is described as:

$$\frac{\partial \rho_A}{\partial t} + \frac{\partial(\rho_A u_A)}{\partial x} + \frac{\partial(\rho_A v_A)}{\partial y} + \frac{\partial(\rho_A w_A)}{\partial z} = \dot{\omega}_A \tag{94}$$

This equation can be written in vector format for simplicity as:

$$\frac{\partial \rho_A}{\partial t} + \nabla \cdot (\rho_A \mathbf{V}_A) = \dot{\omega}_A \tag{95}$$

where the vector $\mathbf{V}_A$ is equal to $[u_A \ v_A \ w_A]$.

In this equation, the mass flux rate of individual species is the product of the specific density of that species and velocity in the given direction. However, the velocity of individual species depends not only on the bulk velocity of the flow, but also on concentration gradients. If there is a difference in concentration of species at various points across the flow, the species will move from regions of high concentrations to that of low concentrations. This motion is analogous to the phenomena of heat conduction from high temperatures to low temperatures. The velocity induced by a concentration gradient is called the diffusion velocity.

Consider a multicomponent system with different species in the mixture moving with different velocities in different directions. As shown in Fig. 6, **V** represents the mass averaged bulk velocity of the flow ($V = ui + vj + wk$) and **V*** represents the molar averaged flow velocity. Both **V** and **V*** will differ in magnitude and direction as they contain unique weighing factors. Letting V_A be the velocity of species A, this value is independent on the molar weight or mass of the species A as it represents only that species. The difference between velocity of individual species and the mass averaged velocity is called the mass diffusion velocity of that species. In this case, it is υ_A. Similarly, the difference between individual species velocity and the molar averaged velocity of the flow is called the molar diffusion velocity υ_A^*.

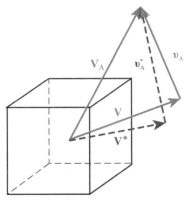

Fig. 6. Diagram of mass and molar averaged bulk velocity vectors.

Therefore, the velocity of individual species A is given as:

$$V_A = V + \upsilon_A \tag{96}$$

Hence, the conservation of species equals:

$$\frac{\partial \rho_A}{\partial t} + \nabla \cdot (\rho_A V) + \nabla \cdot (\rho_A \upsilon_A) = \dot{\omega}_A \tag{97}$$

Adolf Fick investigated the diffusion velocity in the above equation in detail. He explained that υ_A could be expressed using diffusion coefficients (δ). The use of this coefficient in the mass flux expression was later called Fick's law of diffusion and is written in mathematical format as:

$$\rho_A \upsilon_A = -\rho_A \delta_{Am} \nabla (\ln Y_A) \tag{98}$$

where the mass fraction is given by

$$Y_A = \frac{\rho_A}{\rho} \tag{99}$$

with the m subscript on the diffusion coefficient indicating the value of species A with respect to the entire mixture as described later.

Substituting this expression back into Fick's law of diffusion:

$$\rho_A \mathbf{v}_A = -\rho_A \delta_{Am} \nabla\left(\ln\frac{\rho_A}{\rho}\right) \tag{100}$$

solving for the logarithmic derivative,

$$(e.g., \quad \frac{\partial[\ln(\rho_A/\rho)]}{\partial x} = \frac{\rho_A}{\rho}\frac{\partial(\rho/\rho_A)}{\partial x}),$$

$$\rho_A \mathbf{v}_A = -\rho\delta_{Am}\nabla\left(\frac{\rho_A}{\rho}\right) \tag{101}$$

and substituting equations (99) and (101) into the conservation of species, equation (97), recovers:

$$\frac{\partial(\rho Y_A)}{\partial t} + \nabla\cdot(\rho Y_A \mathbf{V}) - \nabla\cdot(\rho\delta_{Am}\nabla Y_A) = \dot{\omega}_A \tag{102}$$

The first two components on the left hand side of equation (102) can be expanded to obtain,

$$\rho\frac{\partial Y_A}{\partial t} + \rho\mathbf{V}\nabla\cdot Y_A + Y_A\left[\underbrace{\frac{\partial\rho}{\partial t} + \nabla\cdot(\rho\mathbf{V})}_{0}\right] - \nabla\cdot(\rho\delta_{Am}\nabla Y_A) = \dot{\omega}_A \tag{103}$$

and conservation of mass requires the term in square brackets of equation (103) to be zero so that:

$$\rho\frac{\partial Y_A}{\partial t} + \rho\mathbf{V}\nabla\cdot Y_A - \nabla\cdot(\rho\delta_{Am}\nabla Y_A) = \dot{\omega}_A \tag{104}$$

This equation is further modified as per the assumptions of dynamic incompressibility. Since the mass fraction of the species is related to the molar concentration of the species by:

$$Y_A = \frac{\bar{C}_A W_A}{\rho} \tag{105}$$

where W_A is the molecular weight of species A, the conservation of species equation for dynamic incompressible flow is:

$$\frac{\partial\bar{C}_A}{\partial t} + \mathbf{V}\nabla\cdot\bar{C}_A - \nabla\cdot(\delta_{Am}\nabla\bar{C}_A) = \frac{\dot{\omega}_A}{W_A} \tag{106}$$

Fick's law expresses the diffusion velocities in terms of diffusion coefficients. However, calculation of these coefficients has always been subject to interpretation in the literature. The standard method of calculating these values is a three-step process. First, the diffusion coefficient for one species is calculated as in a binary mixture consisting of two gases with one gas is held as the base. The binary diffusion coefficient is then calculated for all other gases in the mixture, keeping the base gas constant. Then, the diffusion coefficient of the base gas is calculated for the mixture from all of the calculated binary diffusion coefficients. This method is then repeated for all the gases in the mixture keeping one gas as the base each time.

Many ways have been proposed to calculate the binary diffusion coefficient in the mentioned procedure (Perry & Green, 1984; Cussler, 1997; Kee et al., 2003; Kuo, 2005). After studying the referenced literature, the authors feel that binary diffusion between two species (A and B here) is best calculated by the following (Cussler, 1997; Kee et al., 2003):

$$\delta_{AB} = \frac{0.00186T^{1.5}\left(\dfrac{1}{W_A} + \dfrac{1}{W_B}\right)^{0.5}}{p\sigma_{AB}^2\Omega} \tag{107}$$

Equation (107) assumes all gases to be non polar, and although values of σ (Collision diameter of the molecule) and Ω (dimensionless energy integral based on temperature and Boltzmann constant) are not available for all gases. However, this equation provides for a relatively high accuracy, within an eight percent error range with experimental data (Cussler, 1997). This accuracy is appreciable as it is the closest any equation predicting binary diffusion coefficient can get. Even the commercial software program, transport CHEMKIN, supports this same equation for calculating these values (Reaction Design, 2003). From these binary coefficients, the mixture averaged diffusion coefficients can be calculated. The most accurate method to calculate these coefficients is to use a full multi-component system that involves inverting an L by L matrix, where L is number of species (Wilke, 1950; Perry & Green, 1984; Cussler, 1997; Kee et al., 2003). However, this method is computationally expensive and not required in most numerical models. As a result, most researchers use approximate formulas as follows.

When the mass diffusion velocity is given as a function of mass fractions,

$$\mathbf{v}_A = \frac{-1}{Y_A}\delta_{Am}\nabla Y_A \tag{108}$$

The mixture-averaged diffusion coefficient is represented as:

$$\frac{1}{\delta_{Am}} = \sum_{j\neq A}^{L}\frac{X_A}{\delta_{Aj}} + \frac{X_A}{1-Y_j}\sum_{j\neq A}^{L}\frac{Y_j}{\delta_{Aj}} \tag{109}$$

However, when the mass diffusion velocity is written using mole fractions (or molar concentrations),

$$\mathbf{v}_A = \frac{-1}{X_A}\delta_{Am}'\nabla X_A \tag{110}$$

The mixture-averaged diffusion coefficient is now written as:

$$\delta_{Am}' = \frac{1-Y_A}{\displaystyle\sum_{j\neq A}^{L}\frac{X_j}{\delta_{Aj}}} \tag{111}$$

The reason for the two different methodologies in calculating the diffusion coefficient relates to the solution of the mole or mass fraction version of the governing equation of chemical species. In specific, since mole and mass fractions are related by the molecular weight of the species, diffusion coefficients must be calculated differently in order to take this weighting factor into account. For more information on this topic, please consult the efforts of Kee et al. (Kee et al., 2003).

8. Failure of dynamic incompressibility

As discussed in the Mach Number Criterion section, when the pressure and temperature are not directly related to the density of the medium, one may simplify the governing equations via dynamic incompressibility (Kee et al., 2003). For catalytic exhaust aftertreatment modeling, a previous effort by the first author shows that traditional catalyst model equations assume dynamic incompressibility in order to simplify the solution technique by decoupling the species and energy equations (Depcik & Assanis, 2005). This model, represented by Figure 7, simulates an open channel interacting with the surface of the catalyst (washcoat impregnated with catalytic materials) by virtue of literature derived source terms, similar to equation (91), that model the interactions within the boundary layer. Use of dynamic incompressibility for simulating the gas in the open channel increases the computational speed of the model in order to make it suitable for transient regulatory tests and kinetic constant optimization.

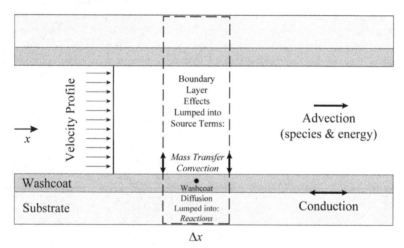

Fig. 7. One-dimensional catalytic exhaust aftertreatment model description.

Since the laminar channels of a monolithic catalyst do not contribute significantly to the pressure drop, the pressure throughout the catalyst is nearly constant. However, when a temperature profile exists, according to the ideal gas law the density must change along the monolith. This was stated previously by Byrne and Norbury (Byrne & Norbury, 1993) where they assert that the variation of gas density with temperature suggests that the flow should be treated as compressible for aftertreatment modeling; the assertion was further verified by Depcik et al. (Depcik et al., 2010). Compressibility leads to variations in velocity of the gas and a decreasing residence time of the gas in the catalyst.

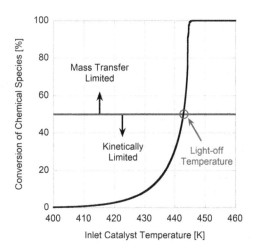

Fig. 8. Example of light-off temperature of a catalyst as indicated by the 50% conversion of the chemical species of interest.

One significant goal of the catalyst model is to simulate a "light-off event" when the catalyst transitions from kinetic to mass transfer regimes as illustrated in Figure 8. In other words, when the temperature is low, the reactions taking place on the surface are slow and there is little conversion of species across the catalyst (chemical kinetic limited). As the temperature increases, the reactions increase in magnitude until they hit a certain activation energy that causes an exponential jump in the magnitude of the conversion (50% conversion of inlet species is typically considered the light-off point). As a result, during a light-off experiment the temperature of the catalyst can change dramatically if a significant amount of exothermic reactions takes place. Hence, from a modeling standpoint, it becomes erroneous to employ dynamic incompressibility at this point. However, many researchers neglect this fact and continue to simulate the gas dynamics in this manner.

After light-off of a catalyst, it is commonplace to assume the reaction is now mass transfer limited and not kinetic limited. Hence, once entering the catalyst, all species will be converted. From a regulatory standpoint, modeling the catalyst after light-off is virtually unnecessary because most emissions occur in the cold start phase when temperatures and reaction speed is low (Koltsakis & Stamatelos, 1997). Therefore, during the times the model is most needed, pressure and temperature through the catalyst are relatively uniform; i.e. the use of dynamic incompressibility is valid.

As a result, it is important to understand that the assumption of dynamic incompressibility fails under situations that involve a significant pressure drop or temperature change. This failure occurs because the fluid is no longer held constant to the ideal gas law and the density remains independent of changes in pressure or temperature. However, by understanding the application of the model within certain parameters, it is possible to utilize dynamic incompressibility in a safe manner.

9. Reaction rate usage

One item that requires further discussion within the catalytic exhaust aftertreatment community is the use of mole fractions for reactions rates; e.g. the right hand side of equation (104). It must be stressed that reactions rates depend on concentration in principle. In the general case when the operation pressure can differ, the more suitable choice for formulating reactions rates is via concentration rate expressions because they inherently take into account changes in density. For example, consider a case of catalytic reactor working at one atmosphere. If reaction rates are expressed in terms of mole fractions, the reaction rate does not respond to the change of pressure. One obtains the same production rates even for a vanishing pressure. When the pressure is lowered, the subsequent production of species is slower because of a less dense mixture. Thus, a model not limited to a singular pressure must be based on concentration and take into account density variations; i.e. similar temperatures at different pressures would result in different densities but could have the same mole fractions. In other words, a model based on concentration can work for different operational pressures; whereas, a model based on mole fractions is fixed for one pressure.

Since pressure changes along the monolith are small and most automotive catalysts work at similar pressures, the use of mole fraction-based expressions does not cause significant problems, but it is important for the reader to understand the implication of doing so. If the reaction rate is written using concentration-based reaction expressions, the results will be significantly different because density is constant and does not respond to the temperature change, as discussed in the previous section; hence, reaction rates will be unique at each cell as axial temperature fluctuations change the calculated rate.

For a more in-depth discussion of this topic with respect to catalytic exhaust aftertreatment devices, please consult reference (Depcik et al., 2010). For simplicity, an example is given here for homogeneous (gas phase) reactions in order to demonstrate the concept. The right hand side of the compressible version of the chemical species equation (104), when homogeneous reactions occur, should be written in terms of the gas concentration

$$\dot{\omega}_A = A\exp\left(\frac{-E}{R_u T}\right)C_{CO} \tag{112}$$

where equation (112) expresses a linear reaction rate dependant on the species carbon monoxide according to the traditional Arrhenius rate expression.

However, when the dynamic incompressible version is used via equation (106)

$$\frac{\dot{\omega}_A}{W_A} = A\exp\left(\frac{-E}{R_u T}\right)X_{CO} \tag{113}$$

mole fractions must be used. The pre-exponential components (A) in both expressions can be equated at specific pressures and temperatures via density.

10. Conclusion

Dynamically incompressible flow is a convenient simplification of the governing equations of fluid mechanics in order to solve for flow conditions below a Mach number of 0.3. This assumption decouples the equations, because the fluid properties are no longer subject to the constraint of the ideal gas law. Consequently, the equations can be solved in a reduced computational time. However, when pressure and temperature change significantly across the computational domain, the assumption of dynamically incompressible flow falters. In order to aid in the development of the reader, this chapter provided the proper background into this computational methodology and illustrated the correct tactic of computing reaction rates for the governing equation of chemical species.

11. References

Anderson, J. J. (1995). *Computational Fluid Dynamics: The Basics with Applications* (1st Ed), McGraw-Hill, 0070016852, New York.

Byrne, H. and J. Norbury (1993). Mathematical Modelling of Catalytic Converters. *Math. Engng. Ind.* vol. 4, no. 1, pp. 27-48.

Cussler, E. L. (1997). *Diffusion: Mass Transfer in Fluid Systems* (2nd Ed), Cambridge University Press, 0521450780, New York.

Depcik, C. and D. Assanis (2005). One-Dimensional Automotive Catalyst Modeling. *Progress in Energy and Combustion Science*, vol. 31, no. 4, pp. 308-369.

Depcik, C., A. Kobiera, et al. (2010). Influence of Density Variation on One-Dimensional Modeling of Exhaust Assisted Catalytic Fuel Reforming. *Heat Transfer Engineering*, vol. 31, no. 13, pp. 1098-1113.

Fox, R., A. McDonald, et al. (2004). *Introduction to Fluid Mechanics* (6th Ed), John Wiley & Sons, Inc., 0471202312, Hoboken, New Jersey.

Gad-el-Hak, M. (1995). Questions in Fluid Mechanics: Stokes' Hypothesis for a Newtonian, Isotropic Fluid. *Journal of Fluids Engineering*, vol. 117, no. 1, p. 3-5.

Graetzel, M. and P. Infelta (2002). *The Bases of Chemical Thermodynamics: Vol. 1*, Universal-Publishers, 1581127723.

Honerkamp, J. (2002). *Statistical Physics: An Advanced Approach with Applications* (2nd), Springer, 3540430202, Berlin.

Kee, R. J., M. E. Coltrin, et al. (2003). *Chemically Reacting Flow: Theory and Practice* (1st Ed), John Wiley & Sons, Inc., 0472361793, Hoboken, New Jersey.

Koltsakis, G. C. and A. M. Stamatelos (1997). Catalytic Automotive Exhaust Aftertreatment. *Progress in Energy and Combustion Science*, vol. 23, no. 1, p. 1-39.

Kreith, F., R. M. Manglik, et al. (2010). *Principles of Heat Transfer* (7th Ed), Cengage Learning, Inc., 0495667706.

Kundu, P. K. and I. Cohen (2010). *Fluid Mechanics* (4th Ed), Academic Press, 0123813999, New York.

Kuo, K. K. (2005). *Principles of Combustion* (2nd Ed), John Wiley & Sons, Inc., 0471046892, Hoboken, New Jersey.

Panton, R. L. (2005). *Incompressible Flow* (3rd Ed), John Wiley & Sons, Inc., 047126122X, Hoboken, New Jersey.

Perry, R. and D. Green (1984). *Perry's Chemical Engineering Handbook* (6th Ed), McGraw-Hill, Inc., 0070494797, New York.

Reaction Design (2003). *Transport Core Utility Manual*. Available from: http://eccentric.mae.cornell.edu/~laniu/MAE643/CHEMKIN3.7/transport.pdf

Roy, S. K. (2001). *Thermal Physics and Statistical Mechanics*, New Age International Publishers, 81-224-1302-1, New Delhi.

Schlichting, H. and K. Gersten (2000). *Boundary Layer Theory* (8th Ed), Springer, 3540662707, Berlin.

Turcotte, D. L. and G. Schubert (2002). *Geodynamics* (2nd), Cambridge University Press, 0521661862, New York.

Vardi, J. and W. F. Biller (1968). Thermal Behavior of an Exhaust Gas Catalytic Converter. *Industrial and Engineering Chemistry - Process Design and Development*, vol. 7, no. 1, p. 83-90.

White, F. (2003). *Fluid Mechanics* (5th Ed), McGraw-Hill, 0072402172, Boston.

White, F. (2006). *Viscous Fluid Flow* (3rd Ed), McGraw-Hill, 0072402318, New York.

Wilke, C. R. (1950). Diffusional Properties of Multicomponent Gases. *Chemical Engineering Science*, vol. 46, no. 2, p. 95-104.

6

Modification of the Charnock Wind Stress Formula to Include the Effects of Free Convection and Swell

C.H. Huang

U.S. Department of the Interior,
Bureau of Ocean Energy Management, Regulation, and Enforcement,
USA

1. Introduction

The estimate of the surface fluxes of momentum, sensible heat, and water vapor over the ocean is important in numerical modeling for weather forecasting, air quality modeling, environmental impact assessment, and climate modeling, which, in particular, requires more accurate flux calculations. The accuracy of flux calculations depends on an important parameter: the surface roughness length. More than five decades ago Charnock (1955) formulated the surface roughness length over the ocean, relating the sea-surface roughness to wind stress on the basis of dimensional argument. Even now the Charnock wind stress formula is widely used. However, in the free convection limit, the horizontal mean wind speed and the wind stress approach zero, therefore the friction velocity also approaches zero; in this condition, the Charnock wind stress formula is no longer applicable and the Monin-Obukhov similarity theory (MOST) has a singularity; MOST relies on the dimensional analysis of relevant key parameters that characterize the flow behavior and the structure of turbulence in the atmospheric surface layer. Thus, in the conditions of free convection, the traditional MOST is no longer valid, because the surface roughness length also approaches zero. In numerical modeling, the surface fluxes, wind, and temperature in the surface layer are important, because they are used as lower boundary conditions.

In the Charnock formulation of surface roughness, two important physical processes were not considered: one is the free convection and the other is the swell. Swell is the long gravity waves generated from distant storms, which modulate the short gravity waves. To date the formulation of the effect of swell on the aerodynamic roughness length has not been theoretically developed. Thus, the purpose of this study is to modify the Charnock wind stress relation to include the effects of the free convection and the swell on the surface roughness length. A formula of the surface roughness length that relates to the convective velocity is proposed by Abdella and D'Alessio (2003). However, in this study, the alternative approaches, which are much simpler than the Abdella and D'Alessio derivation, are used to derive the surface roughness length, which takes into account the effects of the free convection. The convective process is important in affecting the air-sea transfer of momentum, sensible heat, and water vapor, especially at very light winds. Businger (1973) suggested that under the conditions of free convection and in the absence of the horizontal mean wind, averaged over the horizontal area, a local wind profile still exists. This wind profile is generated by the convective circulations in the atmospheric boundary layer and, hence, we can observe a wind stress and a shear production of turbulent energy near the surface. As a consequence of the convective circulations, it will induce the "virtual" horizontal stress, the so-called convection-induced stress, which contributes to the increase of the surface roughness. Figure 1 illustrates the concept of convective circulations over the ocean.

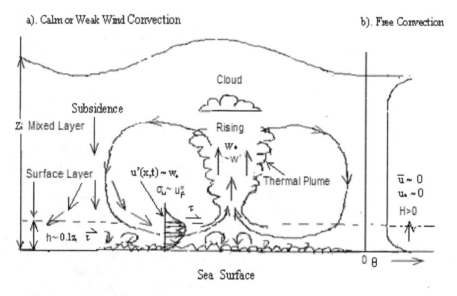

Fig. 1. The figure shows the large scale circulations associated with the convective activities. a. Near Calm or weak wind convection, b. Free Convection. Where z_i is the mixed-layer height, h the surface layer (inner layer), $\bar{u}$ the mean wind speed, u_* the friction velocity, w_* the convective velocity, τ the wind stress, θ the potential temperature, H the sensible heat flux, σ_u the standard deviation of the horizontal mean wind speed, u_{*f} the induced friction velocity, and $u'(x, t)$ the horizontal velocity fluctuations induced by large eddies in connection with the large scale convective circulations.

In free convection, the velocity should be scaled by the convective velocity as proposed by Deardorff (1970) rather than by the friction velocity. The convective velocity is an important scaling parameter in the study of the atmospheric convection. Therefore, in this paper, we introduce a new velocity scale that is a combination of the friction velocity and the convective velocity. This velocity scale is equal to the total friction velocity. With the use of this velocity scale to obtain the surface roughness length, the singularity in the traditional MOST is avoided. In general, the surface roughness length also depends on the sea state and can be expressed in terms of friction velocity, convective velocity, air viscosity, and wave age. The effect of the wave-induced stress or swell on the surface roughness is also investigated in this paper.

In the free convection limit, Abdella and D'Alessio (2003) found that the Charnock relation substantially underestimates the value of the momentum roughness. To remedy this, they proposed a sea surface roughness formula for z_o by including a convective velocity in the Charnock relation as

$$z_o = \frac{\alpha}{g}\left(u_*^2 + \gamma_a w_*^2\right),$$ (1)

where the parameter a is the Charnock constant, which also depends on the wave age, g is the acceleration of gravity, u_* is the friction velocity (The squared value of friction velocity is related to wind stress, $\tau/\rho_a = u_*^2$, where ρ_a is the density of air), w_* is the convective velocity (as defined below in Eq. (27); the convection velocity is also used in the literature), and γ_a is an empirical constant. Eq. (1) reduces to the Charnock relationship if the convective velocity is neglected.

In this paper, we propose an alternative approach for the parameterization of roughness length, z_o, for flow over the sea in forced and free convection (forced convection is without buoyancy forces), which is expressed as

$$z_o = \frac{\alpha}{g}\left(u_*^3 + \gamma w_*^3\right)^{\frac{2}{3}},$$ (2)

where γ is an empirical constant. Equation (2) is similar to that of the roughness length proposed by Abdella and D'Alessio (2003) as shown in Eq. (1). However, in the limiting conditions Eq. (1) and Eq. (2) are identical.

In this paper, first we provide necessary background information about MOST in Section 2 and about the bulk formulations for the surface fluxes of momentum, heat, and water vapor in Section 3. Then, two alternative approaches are used to derive the surface roughness length as given in Eqs. (1) and (2) in Section 4. The first approach is based on the Prandtl mixing length theory and the standard deviation of the vertical velocity component to derive a new velocity scale that is applicable to the conditions of free convection. The second approach is to use the standard deviations of the horizontal velocity fluctuations induced by large eddies in connection with the large scale convective circulations to derive the relationship between surface roughness and the convective velocity, which is consistent with the concept of gustiness proposed by Businger (1973) (also see Schumann, 1988; Sykes et al., 1993).

The swell may also alter the surface roughness. Thus, in this paper, the Charnock wind stress relation including the effect of swell is derived theoretically for the first time.

The background information about MOST is provided in the following section.

2. The Monin-Obukhov similarity theory

The Monin-Obukhov similarity theory is based on dimensional analysis, which basically states that the flow is in quasi steady state over the horizontally homogeneous surface. And the vertical profiles of the horizontal mean wind and temperature, and the characteristics of turbulence in the atmospheric surface layer can be described as a universal function of relevant parameters, including the height above the surface, the surface wind stress, the buoyancy parameter, and the surface heat flux. In the atmospheric surface layer, the vertical variations of surface fluxes are within 10%. This surface layer is also called the constant stress layer or constant flux layer; the wind stress is aligned with the direction of the horizontal mean wind.

According to the well-known Monin-Obukhov similarity theory (MOST) (Monin-Obukhov, 1954), the non-dimensional profiles of wind shear, ϕ_m, temperature gradient, ϕ_h, and moisture gradient, ϕ_q, are expressed as:

$$\frac{kz}{u_*}\frac{\partial \bar{u}}{\partial z} = \phi_m\left(\frac{z}{L}\right),$$

(3)

$$\frac{kz}{\theta_*}\frac{\partial \bar{\theta}}{\partial z} = \phi_h\left(\frac{z}{L}\right),$$

(4)

$$\frac{kz}{q_*}\frac{\partial \bar{q}}{\partial z} = \phi_q\left(\frac{z}{L}\right).$$

(5)

In the above equations, it states that in the atmospheric surface layer, the non-dimensional wind shear, and temperature and moisture gradients can be expressed as a universal function of the atmospheric stability parameter, z/L, where L is the Monin-Obukhov stability length (as defined in Eq. (8-1)). And where θ_* and q_* are the temperature and moisture scales, respectively, defined as

$$\theta_* = -\frac{\overline{w'\theta'}}{u_*}$$

(6)

$$q_* = -\frac{\overline{w'q'}}{u_*}$$

(7)

Where $\bar{u}$, is the horizontal mean wind speed; $\bar{\theta}$, is the mean potential temperature; $\bar{q}$, is the mean specific humidity; k is the von Kármán constant (k=0.4); $\overline{w'\theta'}$ is the surface kinematic heat flux; $\overline{w'q'}$ is the moisture flux, and L is the Monin-Obukhov stability length (also called Obukhov length (Obukhov,1946)) defined as

$$L = -\frac{\theta_v u_*^3}{kg\left(\overline{w'\theta_v'}\right)},$$

(8-1)

or

$$L = -\frac{T(1+0.61q)u_*^2}{kg[\theta_* + 0.61Tq_*]},$$

(8-2)

where T is the temperature, and θ_v is the virtual temperature. Eq. (8-1) and Eq.(8-2) are equal to each other; Eq.(8-2) is expressed in terms of the scaling parameters, , u_*, θ_*, and q_*.

For practical applications, we assume the following flux-profile relations (see Huang, 1979):

$$\phi_h = \phi_q = \phi_m^2, \quad z/L < 0 \text{ for unstable conditions}$$

(9)

where $\phi_h(z/L)$ is the non-dimensional function of temperature gradient.

Based on various meteorological experiments, different forms of the universal function for $\phi_m(z/L)$ have been proposed. For example, the universal function for the non-dimensional wind shear $\phi_m(z/L)$ empirically obtained from the observational data (Dyer, 1974; also see Huang, 1979) can be expressed as:

$$\phi_m\left(\frac{z}{L}\right) = 1 + \alpha_1 \frac{z}{L} \qquad z/L \geq 0 \text{ for stable conditions} \tag{10-1}$$

$$\phi_m\left(\frac{z}{L}\right) = \left(1 - \alpha_2 \left(\frac{z}{L}\right)\right)^{-\beta_1}, \quad z/L < 0 \text{ for unstable conditions} \tag{10-2}$$

where the values of $a_1 = 5$, $a_2 = 16$ and $\beta_1 = 1/4$ are recommended by Dyer (1974) for unstable conditions. Other values of $a_2 = 7$ and $\beta_1 = 1/3$ for unstable conditions are given by Troen and Mahrt (1986) and others.

In general, the function of $\phi_h(z/L)$ can be expressed as (Hogstrom, 1998; Wilson, 2001):

$$\phi_h = \text{Pr}_t\left(1 + \beta_h \frac{z}{L}\right) \quad z/L \geq 0 \text{ for stable conditions} \tag{11-1}$$

$$= \text{Pr}_t \quad z/L = 0 \text{ for neutral conditions} \tag{11-2}$$

$$= \text{Pr}_t\left(1 - \gamma_h \frac{z}{L}\right)^{-\frac{1}{2}} \quad z/L < 0 \text{ for unstable conditions} \tag{11-3}$$

Where β_h and γ_h are empirical constants and the values of β_h and γ_h are equal to 5 and 16, respectively. Pr_t is the turbulent Prandtl number and is defined as

$$\text{Pr}_t = \frac{K_m}{K_h}. \tag{11-4}$$

Where K_m and K_h are exchange coefficients (or eddy diffusivities) for momentum and heat, respectively. The turbulent Prandtl number for neutral stability, Pr_t, is equal to 0.95 .

3. Fluxes of momentum, heat, and moisture

The standard bulk formulations for the surface fluxes of momentum (τ), heat (H), and moisture (E) can be expressed as

$$\tau = \rho_a C_d \left(u_s - u_r\right)^2 = \rho_a u_*^2, \tag{12}$$

$$H = \rho_a c_{pa} C_h u_r \left(T_s - \theta\right) = -\rho_a c_{pa} u_* \theta_* \tag{13}$$

$$E = \rho_a L_e C_e u_r \left(q_s - q\right) = -\rho_a u_* q_*. \tag{14}$$

Where ρ_a is the air density, c_{pa} the specific heat of air at constant pressure, u_r the mean wind speed at the reference height z_r above the sea surface, u_s the speed of surface current, T_s the skin temperature, q_s the saturation specific humidity at the sea surface, and L_e the latent heat of water vapor. The C_d, C_h, and C_e are the bulk transfer coefficients for wind stress or momentum, sensible heat, and latent heat, respectively. The typical values of the bulk transfer coefficients are about 0.001 at the reference height of about 10 m. For the ease of computation, these transfer coefficients can be further partitioned into individual components and written as analytical functions that depend on the atmospheric stability parameter (z/L) as

$$C_d = C_d^{1/2}C_d^{1/2} = \frac{C_{dn}}{\left[1 - \frac{C_{dn}^{1/2}}{k}\psi_m\left(\frac{z}{L}\right)\right]^2} \, , \tag{15}$$

$$C_h = C_d^{1/2}C_T^{1/2} = \frac{C_{dn}^{1/2}C_{hn}^{1/2}}{\left[1 - \frac{C_{dn}^{1/2}}{k}\psi_m\left(\frac{z}{L}\right)\right]\left[1 - \frac{C_{hn}^{1/2}}{k}\psi_h\left(\frac{z}{L}\right)\right]} \, , \tag{16}$$

$$C_e = C_d^{1/2}C_q^{1/2} = \frac{C_{dn}^{1/2}C_{hn}^{1/2}}{\left[1 - \frac{C_{dn}^{1/2}}{k}\psi_m\left(\frac{z}{L}\right)\right]\left[1 - \frac{C_{qn}^{1/2}}{k}\psi_q\left(\frac{z}{L}\right)\right]} \, . \tag{17}$$

These components can be expressed on the basis of the Monin-Obukhov similarity theory as functions of reference height, z, surface roughness length (z_o for momentum, z_{oT}, for heat, and z_{oq} for water vapor), and the Monin-Obukhov length, L, defined in Eq. (8). For more detailed information on the functional expressions of the bulk transfer coefficient, refer to texts such as Garratt (1992) or Stull (1997). Where $\psi_m(z/L)$, $\psi_h(z/L)$, and $\psi_q(z/L)$ are the similarity functions, which are given in Eq. (19) to (21) below. It is usually assumed that there is similarity between water vapor and heat transfer, which implies that $\psi_h(z/L) = \psi_q(z/L)$.

For example, the bulk transfer coefficient of momentum or the wind profile can be expressed as (see Paulson, 1970; Huang 1979)

$$C_d \equiv \left(\frac{u_*}{u}\right)^2 = \frac{k^2}{\left(\ln\left(\frac{z}{z_o}\right) - \psi_m\right)^2} \, . \tag{18}$$

Where the similarity function $\psi_m(z/L)$ in Eq. (15) or (18) is given as

$$\psi_m = \psi_h = -5\frac{z}{L} \quad \text{for } z/L \geq 0 \tag{19}$$

$$\psi_m = 2\ln\left(\frac{1+x}{2}\right) + \ln\left(\frac{1+x^2}{2}\right) - 2\tan^{-1}(x) + \frac{\pi}{2} \quad \text{for } z/L < 0 \tag{20}$$

and the function $\psi_h(z/L)$ in Eq. (16) is given as

$$\psi_h = 2\ln\left(\frac{1+x^2}{2}\right) \quad \text{for } z/L<0 \tag{21}$$

where

$$x = \left(1 - 16\frac{z}{L}\right)^{1/4} \quad \text{for } z/L < 0$$

Under neutral conditions, that is when z/L approaches 0, the transfer coefficients in Eqs. (15) – (17) become

$$C_{dn}^{1/2} = \frac{k}{\ln\left(\dfrac{z_r}{z_o}\right)}, \tag{22}$$

$$C_{hn}^{1/2} = \frac{k}{\ln\left(\dfrac{z_r}{z_{oT}}\right)}, \tag{23}$$

$$C_{qn}^{1/2} = \frac{k}{\ln\left(\dfrac{z_r}{z_{oq}}\right)}. \tag{24}$$

Where z_{oT} and z_{oq} are roughness lengths for heat and water vapor, respectively.

4. Relationship between the friction and convective velocity

Two approaches are used here to derive the relationship between the induced friction velocity and the convective velocity. These approaches are based on the Prandtl mixing length theory and the turbulent velocity fluctuations, which are described in the following sections.

4.1 Prandtl mixing length theory and velocity scale
In this section, we use the Prandtl mixing length theory and the standard deviation of the vertical turbulent velocity to derive the surface roughness length that includes the effect of free convection on the surface roughness length. The Prandtl mixing length theory has also been used to derive the Abdella and D'Alessio formula in Eq. (2) (also see Huang, 2009).

According to MOST, the standard deviation of the vertical turbulent velocity component (w'), $\sigma_w = \sqrt{(w')^2}$, in the surface layer can be described by:

$$\frac{\sigma_w}{u_*} = \phi_w\left(\frac{z}{L}\right), \tag{25}$$

In Eq. (25), ϕ_w is a non-dimensional function, which depends only on the height, z, and the Monin-Obukhov stability length, L.

The similarity function for the vertical turbulent velocity postulated by Wyngaard and Cote (1974) and suggested by Panofsky et al. (1977) can be expressed as

$$\frac{\sigma_w}{u_*} = 1.25\left(1 - 3\frac{z}{L}\right)^{\frac{1}{3}}. \tag{26}$$

Equation (26) is based on the similarity theory and observational data. Under the conditions of free convection, buoyancy is the driving force, which, through the free convection in connection with the large-scale convective circulation, induces the horizontal motions (see Fig. 1).

The following, through the Prandtl mixing length theory, shows that the standard deviation of the vertical turbulent velocity is related to the total horizontal wind stress.

The convective velocity scale, w_*, is expressed by Deardorff (1970) as

$$w_* \equiv \left(\frac{g}{T}H_o z_i\right)^{\frac{1}{3}} = \left(-\frac{z_i}{kL}\right)^{\frac{1}{3}}.u_* . \tag{27}$$

Where T is the temperature of air, H_o is the surface heat flux, and z_i is the depth of the mixed layer. The appropriate scaling parameters for the mixed-layer similarity are the convective velocity, w_*, and the mixing height, z_i.

By using the definition of the convective velocity, w_*, in Eq. (27), Eq. (26) becomes

$$\sigma_w = 1.25\left(u_*^3 + \gamma w_*^3\right)^{\frac{1}{3}}, \tag{28}$$

where γ is an empirical constant. It also can be specified as $\gamma = 3k\varepsilon$ and $\varepsilon = z/z_i$, the ratio of the surface boundary layer to the mixed-layer height, where k is the von Kármán constant and ε is commonly specified to be 0.1 (Troen and Mahrt, 1986).

In the limit of free convection, the friction velocity approaches zero and the component of vertical turbulent velocity, σ_w, in Eq. (28), becomes

$$\sigma_w = \sqrt{w'^2} = 1.25\gamma^{\frac{1}{3}}w_* \approx 0.6w_* . \tag{29}$$

From Eq. (29), we obtain the value of 0.11 for γ. If we set the value of $\varepsilon = 0.1$, then we obtain the value of 0.12 for γ in Eq. (28).

According to the Prandtl mixing length theory, the eddy diffusivity for momentum can be written as

$$K_m = Ckz\sigma_w , \tag{30}$$

where C is a constant to be determined and K_m is related to the wind stress by definition. Substituting Eq. (28) into Eq. (30), we obtain

$$K_m = C'kz\left(u_*^3 + \gamma w_*^3\right)^{\frac{1}{3}}, \tag{31}$$

where C' is a constant. Under the neutral condition, C' is equal to 1. Therefore, Eq. (31) becomes

$$K_m = kz\left(u_*^3 + \gamma w_*^3\right)^{\frac{1}{3}} \qquad . \tag{32}$$

From the above equation, now for convenience, we define a new velocity scale, w_s, as

$$w_s = \left(u_*^3 + \gamma w_*^3\right)^{1/3} . \tag{33}$$

Then Eq. (32) becomes

$$K_m = kz\tilde{u}_* = kzw_s , \tag{34}$$

where $\tilde{u}_*$ is the total friction velocity.
Therefore, from Eqs. (33) and (34), the total stress or the square of friction velocity, $\tilde{u}_*$, can be written as

$$\tilde{u}_*^2 = \left(u_*^3 + \gamma w_*^3\right)^{\frac{2}{3}} . \tag{35}$$

Eq. (35) shows that the free convection represented by the convective velocity, w_*, will induce the surface stress, $\tau = \rho\tilde{u}_*^2$, as proposed by Businger (1973). Eq. (35) can be used to calculate the surface roughness length. The Abdella and D'Alessio derivation of the roughness length formula is much more complicated and involves more assumptions (Abdella and D'Alessio, 2003). The present derivation confirms the Abdella and D'Alessio formulation (see Eq. (1)) and is much simpler than their derivation.
The above approach is consistent with the concept of gustiness produced by large eddies in connection with the convective circulation (see Fig. 1, Schuman, 1988, and Sykes et al. 1993 for details). In the following section, we derive the relationship between the convection-induced wind stress and the convective velocity using this concept.

4.2 Horizontal velocity variance and gustiness
In this section, the horizontal turbulent velocity representing the gustiness is used to obtain the relationship between the convection-induced stress and the convective velocity, which in turn can be used to estimate the surface roughness length.
In convective conditions, it has been found that the gustiness, the large-scale turbulence, will enhance the wave growth and the energy transfer from wind to wave (see Janssen, 1989).
In free convection conditions, the fluctuations of the horizontal wind velocity are controlled primarily by the large-scale convective circulations in the convective boundary layer (CBL) (see Fig. 1). Therefore, the relationship between the standard deviations of the horizontal turbulence components and the convective velocity can be expressed as (e.g., see Fairall et al., 1996)

$$\sigma_u^2 + \sigma_v^2 = \beta^2 w_*^2 , \tag{36}$$

where the gustiness parameter β is an empirical constant, σ_u and σ_v are the standard deviations of horizontal turbulent velocity components, respectively, and $w*$ the convective velocity. A value of β (=1.25) was determined from the Moana Wave measurements of the horizontal velocity variances (Fairall et al. 1996). However, according to Businger (1973), in the limit of free convection, as the mean "friction velocity", $u*$ (or the mean wind stress, $\rho_a u*^2$), and the horizontal mean wind speed, averaged over an area, approach zero, close to the ground, the fluctuations of horizontal velocity induced by the large-scale convective circulations locally promote a convection-induced stress. The standard deviations of horizontal turbulent velocity components that represent the fluctuations of horizontal wind speed will produce the horizontal-induced wind stress. Under convective conditions, the standard deviation of horizontal turbulent velocity is related to the surface friction velocity (see e.g., Monin-Yaglom, 1971; Businger, 1973); therefore, we express the relation between the velocity variances and the convection-induced friction velocity $u*_f$ or the velocity scale ($w_s = u*_f$) as

$$\sigma_u^2 + \sigma_v^2 = \beta_2 u*_f^2 , \tag{37}$$

where β_2 is an empirical constant. For example, under convective conditions, the velocity variances are estimated to be $\sigma_u/u*_f \approx 2.1$ and $\sigma_v/u*_f \approx 2.0$ for the Monin-Obukhov stability parameter of $z/L = -0.6$ (see Monin and Yaglom , 1971); therefore, in this case, we obtain a value of $\beta_2 = 8.41$. This value is used in this study.

From Eqs. (36) and (37), we have the convection-induced stress ($\tau_f = \rho u*_f^2$) as

$$u*_f^2 = \frac{\beta^2}{\beta_2} w_*^2 = \gamma' w_*^2 , \tag{38}$$

where γ' is an empirical constant and is estimated to be $\gamma' = \beta^2/\beta_2 = 0.19$ from Eq. (38), the value of γ' (=$\gamma^{2/3} = 0.23$) has also been estimated above in Section 4.1. Eq. (38) is consistent with the concept of gustiness proposed by Businger (1973) and others (Schumann, 1988; Sykes et al., 1993). Under strong convection conditions, the production of turbulent energy due to buoyancy force is more effective than the shear production of turbulent energy. That is, under free convection, the convective velocity is an important scaling parameter that has been recognized by Deardorff (1970).

5. Effective total stress

Over ocean waves, the effective total stress, τ_{eff}, above the sea surface can be partitioned into four parts: turbulent shear stress, τ_t; wave-induced stress, τ_w; convection-induced stress, τ_f; and viscous stress, τ_v. Thus the effective total stress is defined here as:

$$\tau_{eff}(z, z_i) = \tau_t(z) + \tau_w(z) + \tau_f(z_i) + \tau_v(z) . \tag{39}$$

In the above equation, the effective total stress is a function of height, z, above the sea surface and the mixed layer height, z_i, in the atmospheric boundary layer. The third term on the right-hand side of Eq. (39) has not hitherto been considered. That is under the free

convection and in the absence of swell, τ_{eff} is equal to τ_f (i.e., $\tau_{eff} = \tau_f$). The value of the convection induced stress is equal to $\tau_f/\rho_a = \gamma' w_*^2 = 0.2$ if we set $\gamma' = 0.2$ and $w_* = 1$.
The momentum flux corresponding to each term in Eq. (39) can be written in the form:

$$u_{*eff}^2 = u_{*t}^2 + u_{*w}^2 + u_{*f}^2 + u_{*v}^2 . \tag{40}$$

Omitting the convection-induced stress, τ_f, in Eq. (39), the effective total stress reduces to (see Phillips, 1966, p. 93):

$$\tau(z) = \tau_t(z) + \tau_w(z) + \tau_v(z) . \tag{41}$$

On the basis of the conservation of momentum flux in the marine atmospheric surface layer, the total stress is independent of height, i.e., one assumes that the flow is in a steady state, with horizontal homogeneity in the surface boundary layer (see e.g., Phillips, (1966); Janssen (1991); the WAM model). It is commonly assumed that the total stress (both turbulent and wave momentum and energy) is constant in the atmospheric surface layer, which is usually considered as about the lowest 10 – 30 m above the sea surface.

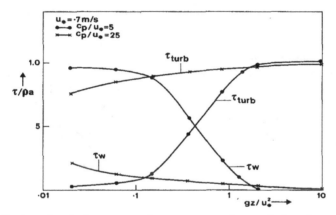

Fig. 2. Distribution of turbulent stress, viscous stress and wave-induced stress as a function of height. The contribution of viscous stress is not shown because it is smaller than 5%. (From Janssen, 1989).

Figure 2 shows the distribution of turbulent stress, viscous stress and wave-induced stress as a function of height, the change of each term in Eq. (41) with height. According to Phillips (1966), the viscous stress can be neglected, except possibly very near the water surface. The wave-induced stress is important only in the wave boundary layer (WBL; see Janssen, 1989). Near the water surface, the wave-induced stress would be significant; however, it decays rapidly with height in the surface layer. Well above the water surface, the wave-induced disturbance vanishes (see Phillips, 1966; Makin and Kudryavtsev, 1999; and Makin, 2008).

6. Modification of the charnock wind stress formula

The surface roughness length, z_o, will be shown to relate to the friction velocity, wave-induced friction velocity, and convective velocity. The well-known Charnock wind stress formula is a relationship between the surface roughness length over the sea and the wind stress or momentum flux; it is expressed as (Charnock, 1955)

$$z_o = \alpha \frac{u_*^2}{g},$$ (42)

where the parameter $a = a_{ch}$ is the Charnock constant with the value ranging from 0.011 (Smith, 1980, 1988) to 0.035 (e.g., see Garratt, 1977; 1992, Table 4.1). Charnock (1955) postulated the roughness-wind stress formulation in Eq. (42) on the basis of a dimensional argument. In the Charnock wind stress relation, Eq. (42), the roughness length depends only on the friction velocity and the gravitational acceleration; the buoyancy effect due to thermal convection is ignored. In this paper, we have shown that under strong convection, the roughness length also depends on the convective velocity (see Section 4).

The dependencies of the surface roughness length on the convective velocity, the wave age, and the effect of the wave-induced stress or swell are investigated in the following sections.

6.1 General Charnock relation and free convection
A general formula for the Charnock relation and the free convection is discussed in this section.

6.1.1 General Charnock relation
In this section, we focus on the effect of free convection on the surface roughness. The general equation for the Charnock relation is obtained by replacing the friction velocity scale (u_*) in the Charnock relation, Eq. (42), with the effective total friction velocity, u^*_{eff}, in Eq. (40) or the vertical velocity scale, w_s, which is a combination of the friction velocity and convective velocity. Therefore, the Charnock formula is also applicable under the conditions of free convection.

Substituting Eq. (40) into Eq. (42) and neglecting the viscous stress for the moment, we have

$$z_o = \frac{\alpha}{\rho_a g} \left(\tau_{eff} \right)$$
$$= \frac{\alpha}{\rho_a g} \left(\tau_t + \tau_w + \tau_f \right),.$$ (43)

where ρ_a is the density of air. Taking the flow over the aerodynamically smooth surface into consideration by including the viscous term, the equation for the surface roughness, Eq. (43), becomes

$$z_o = \frac{\alpha}{\rho_a g} \left(\tau_{eff} \right) + 0.11 \frac{v}{u_*}$$
$$= \frac{\alpha}{\rho_a g} \left(\tau_t + \tau_w + \tau_f \right) + 0.11 \frac{v}{u_*}.$$ (44)

The above equation includes the effect of the wave-induced stress or swell on the surface roughness (see Section 6.3 and 6.4 for more information), and v is the kinematic viscosity (molecular) of air.

The velocity scale as shown in Eq. (33) also includes the effect of buoyancy that will enhance the surface roughness length. By substituting the vertical velocity scale, $w_{s,}$ in Eq. (33) or the total friction velocity in Eq. (35) into the Charnock formula (42) for the friction velocity, u_*, we obtain:

$$z_0 = \frac{\alpha}{g}\left(u_*^3 + \gamma w_*^3\right)^{2/3},$$

(45)

where the value of γ is equal to 0.11. Eq. (45) reduces to the Charnock relation (42) if the free convection term is neglected.

In comparing Eq. (45) with Eq. (43), if we identify the friction velocity in Eq. (45) with the total stress, $\tau/\rho_a = u_*{}^2$ (the friction velocity is related to the wind stress) in Eq. (43) and incorporate the viscous effect, then Eq. (45) or Eq. (44) becomes

$$z_0 = \frac{\alpha}{g}\left(u_*^3 + \gamma w_*^3\right)^{2/3} + 0.11\frac{v}{u_*}$$

$$= \frac{\alpha}{g}\left((u_{*t}^2 \pm u_{*w}^2)^{\frac{3}{2}} + \gamma w_*^3\right)^{2/3} + 0.11\frac{v}{u_*}.$$

(46)

The surface roughness length is dominated by short gravity waves for relatively high wind speed. For light winds, the term 0.11 v/u_* should be added to Eq. (44) or Eq. (46), as suggested by Smith (1988), where $v = 1.5 \times 10^{-5}$ m²/s is the kinematic viscosity of air. This term is for the air flow over an aerodynamically smooth surface. The wave-induced stress usually decreases rapidly with height (see Fig. 2.); therefore, for the height far above the sea surface (above the WBL), $u_* \approx u_{*t}$ (see Phillips, 1966; Makin and Kudryavtsev, 1999; and Makin, 2008). The negative sign in Eq. (46) indicates that the upward transfer of momentum flux from the sea surface to the atmosphere is due to the wave effect. The addition of the viscous term in Eq. (46) is to show that Eq. (46) reduces to Smith's formulation (see Eq. (50) below).

In light wind conditions, swell may influence the roughness length and the momentum transfer. The Charnock parameter in reality is not a constant; it depends on the wave age and may also be influenced by the presence of swell. The effects of the wave age and the wave-induced stress or swell on the Charnock parameter are discussed in Sections 6.2, 6.3, and 6.4, respectively.

6.1.2 Free convection

Equation (46) is a general equation for determining the surface roughness length for the flow over the sea. Thus, we have extended the Charnock wind stress formula to include the conditions of free convection and the wave effect (see Sections 6.3 and 6.4).

In the free convection limit, the horizontal mean wind speed and the shear stress vanish; therefore, the surface roughness length in Eq. (45) becomes

$$z_0 = \frac{\alpha\gamma}{g}w_*^2,$$

(47)

Where $\gamma' = \gamma_a$ is a constant, which is equal to 0.19 to 0.23 (see Section 4). Eq. (47) shows that the surface roughness length is generated under free convection. Eq. (47) has also been proposed by Abdella and D'Alessio (2003); they give a value of 0.15 for γ'.

Here we give two examples to show the effect of free convection on the sea-surface roughness. If we set $\gamma' = 0.23$ and $a = 0.015$, and, in addition, if we assume that $w_* = 1$ m/s, from Eq. (47) we have the surface roughness length $z_o = 3.5x\ 10^{-4}$ m. If we set $w_* = 0.5$ m/s, we have $z_o = 9 \times 10^{-5}$ m .

For illustration, the variation of the surface roughness length with the wind speed under the forced or free convection is plotted in Fig. 3, which shows that the surface roughness length is obtained from the Charnock wind stress formula, omitting the convective term in Eq. (45), and, in the free convection limit as the mean wind speed and shear stress approach zero, the surface roughness is obtained from Eq. (47).

Surface Roughness vs. Wind Speed

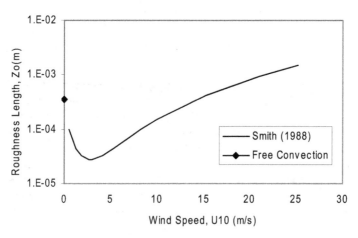

Fig. 3. Variations of the surface roughness over the sea with wind speed at 10 m. The solid line (-) indicates the roughness length obtained from the Charnock relation (see Eq. 50 below). The diamond symbol (♦) shows the roughness length for the free convection; the convective velocity here is set at 1 m/s.

6.2 Effect of wave age

The sea state controls the direction of momentum transfer and also influences the surface roughness. The wave age is often used to characterize the sea state, which is a measure of the sea state for wind wave development. The wave age parameter is defined as c_p/U_{10}, or c_p/u_* (In general the friction velocity, u_* , increases with increasing the wind speed, U_{10}.), where c_p is the phase speed, the speed of wave, and U_{10} is the wind speed at 10 m height above the sea surface, is used as criteria to distinguish the wind sea from the swell. For pure wind sea or for a young wave, the wind speed is larger than the phase speed, i.e., $c_p/U_{10} <$ 1.2, i.e., the wind is moving faster than the wave, while under swell conditions, the waves generated from distant storms travel faster than the wind speed, i.e., $c_p/U_{10} > 1.2$.

In reality, the Charnock constant, a, is not a constant; it also depends on the sea state. The Charnock constant, a, in the roughness length equation, Eq. (45) or Eq. (46), can be modified to better reflect the sea state. The effect of the sea state on the roughness can be modeled by expressing the Charnock constant as a function of the wave age (e.g., Smith et al., 1992) or the wave-induced stress (Janssen, 1989; 1991). A number of authors have suggested that the Charnock parameter a depends on the characteristics of the sea state in the developing stage of wind sea (e.g., Smith et al., 1992), specifically, the wave age, which may be expressed, for example, as

$$\alpha = f\left(\frac{u_*}{c_p}\right) = \alpha_3 \left(\frac{u_*}{c_p}\right)^{\beta_3}, \tag{48}$$

where c_p is the phase speed of the dominant waves and c_p/u_* is the wave age. Young wave has small value of c_p/u_*, while old wave has large value of c_p/u_*. Several authors give different formulations for the function of wave age, $f\ (c_p/u_*)$. Eq. (48) indicates that the young wave has the larger surface roughness, while the older, more mature wave has the smaller surface roughness. Smith et al. (1992) also suggested including the effect of wave age into the Charnock parameter, a. For example, the values of $a_3 = 0.48$ and $\beta_3 = 1$ are given by Smith et al. (1992), while Johnson et al. (1998) give the values of $a_3 = 1.89$ and $\beta_3 = 1.59$. Eq. (46), with the Charnock parameter a replaced by Eq. (48), becomes

$$z_0 = \frac{\alpha_3}{g}\left(\frac{u_*}{c_p}\right)^{\beta_3} \left(u_*^3 + \gamma w_*^3\right)^{2/3} + 0.11\frac{v}{u_*}. \tag{49}$$

Eq. (49) shows that the sea surface roughness length, z_0, is related to the wind stress ($u_* = u_{*t}$), convective velocity, wave age, and air viscosity. Thus, in the free convection limit, the singularity in the Charnock relation is avoided, neglecting the viscous term in Eq. (49). If the effects of free convection and wave age on the surface roughness are neglected, then Eq. (49) becomes

$$z_0 = \frac{\alpha}{g}u_*^2 + 0.11\frac{v}{u_*}. \tag{50}$$

This equation, suggested by Smith (1988), is commonly used to calculate the sea roughness length, z_0, and hence to compute the surface fluxes over the ocean (e.g., Smith, 1988; also see Fairall et al., 1996). For an aerodynamically smooth surface, Eq. (50) reduces to

$$z_0 = 0.11\frac{v}{u_*}. \tag{51}$$

Eq. (51) corresponds to light wind conditions for flow over aerodynamically smooth surface (see Nikuradse, 1933).

6.3 Effect of wave-induced stress
Now, let us take a look at the contribution of the wave-induced stress, τ_w, in Eq. (39) or Eq. (40) to the total stress, τ_{eff}.

For old sea, the magnitude of the wave-induced stress is approximately 10% of the total turbulent stress (see Phillips, 1966; Janssen, 1989). In the atmospheric surface layer (or the WBL) the wave-induced stress usually decays rapidly with height (Janssen, 1989; (see Phillips, 1966; Makin and Kudryavtsev , 1999; and Makin , 2008) (see Section 6.3.1. below for further discussion).

6.3.1 Wind sea and wave-induced stress

The wave-induced stress may influence the surface roughness. To take into consideration the effect of wave-induced stress on the roughness length, Janssen (1989; 1991) suggested modifying the Charnock constant, a by including the wave-induced stress. This approach has been implemented in the wave model (WAM; WAMDI, 1988). The WAM model is widely used in predicting the characteristics of the surface wave.

In the ocean wave modeling, the effect of waves on the surface roughness is usually accommodated through the parameterization of the Charnock "constant" (see Eq. (52) below). The Charnock parameter has been used successfully in the WAM model in the prediction of ocean waves and the effect of wind waves on the transfer of fluxes.

Now, let us consider the effect of the wave-induced stress on the Charnock parameter. Janssen (1991) suggested modifying the Charnock parameter by including the effect of the wave-induced stress on the surface roughness. Thus, Eq. (44) can be written as

$$z_o = \frac{\alpha}{\rho_a g}\left(\tau_t + \tau_f\right) + 0.11\frac{\nu}{u_*}, \tag{52}$$

where

$$\alpha = \frac{\alpha_{ch}}{\sqrt{1-\eta}} \tag{53}$$

and

$$\eta = \frac{\tau_w}{\tau}$$

where a_{ch} is the Charnock constant. The value of a in Eq. (52) is the modified Charnock constant; it includes the effect of the wave-induced stress that depends on the wave age. In the Charnock relation, a is equal to $a_{ch} = 0.0144$ (see Janssen, 1991, p.1634), while in a coupled ocean-wave-atmospheric model, the value of $a_{ch} = 0.01$ is used (WAM; WAMDI, 1988). The WAM model is widely used to forecast the characteristics of ocean wave at meteorological centers around the world.

Substituting Eq. (53) into Eq. (45) or Eq. (46), we obtain

$$z_o = \frac{\alpha_{ch}}{g\sqrt{1-\eta_{eff}}}\left(u_{*t}^3 + \gamma w_{:*}^3\right)^{2/3} + 0.11\frac{\nu}{u_*}, \tag{54}$$

where

$$\eta_{eff} = \frac{\tau_w}{\tau_{eff}}.$$

Under strong convection, the second term on the right-hand side of Eq. (54), the convection-induced stress, is the dominant term. If we neglect the convection-induced stress and the viscous stress, then Eq. (54) becomes

$$z_o = \frac{\alpha_{ch} u_{*t}^2}{g\sqrt{1 - \eta_{eff}}},$$ (55)

where the wave parameter, η, is defined as $\eta_{eff} = \tau_w / \tau_{eff}$ and the effective wave parameter is the ratio of the wave-induced stress, τ_w, to the effective total stress. This wave parameter, η, depends on the wave age and decreases with increasing wave age. The typical value of η is about 10% to 20% (see Phillips, 1966; Phillips, 1977; Janssen, 1989). Phillips (1966) and Janssen (1989) suggested that the total stress for old sea is, in general, $|\tau_w/\tau| \leq 0.2$ and that, with the wind speed less than 2.5 m/s at the height of 10 m, the value of η for old sea is about 10% of the total stress (Phillips, 1966; Janssen, 1989). These values indicate that the effect of wave-induced stress on the value of the Charnock parameter is about 10%; that is, if we assume that, $|\tau_w/\tau| \leq 0.2$, a change of 20% in the wave-induced stress will result in the departure of about 10% from the Charnock constant, a_{ch}.

The direct impact of swell-induced momentum and energy fluxes are confined in the WBL, a thin layer near the water surface for light winds (see Janssen, 1991; Makin and Kudryavtsev, 1999; Makin, 2008). In their theory, the wave boundary layer is in the order of O (1 m), a value less than 10 m. They suggested the height of the WBL is 10 m (Makin and Kudryavtsev, 1999, p. 7615) which is a good estimate (in their Fig. 2, the WBL is much lower and about 1-2 m). The wave-induced stress is exponential decay with height in the WBL, e.g., $f(z) \rightarrow 0$ at $z \rightarrow \delta_w$ (WBL) (see Makin, 2008, p.472; also see the critical layer theory by Miles, 1957).

Thus, the Charnock relation can be considered as a good approximation for moderate wind speeds, that is, $a \approx a_{ch}$, in view of the variability and uncertainty of the data.

In the following section, we consider the effect of swell on the surface roughness.

6.4 Swell effect

Swell is the long gravity waves generated from distant storms. It is common practice to assume that for the air flow over the ocean, under neutral conditions and in the absence of swell, the vertical distribution of the mean wind follows the well-known logarithmic wind profile. Davidson's data were obtained under the influence of heavy swell (Davidson, 1974). Davidson's formulation of the drag coefficient under swell conditions (Davidson, 1974) can be used to transform the wave-dependent formula of the drag coefficient into the following form:

$$Cd = \frac{Cd_o}{\left(1 + \beta'\left(\frac{c_p}{u_*} - \alpha_o\right)\right)^2},$$ (56-1)

where

$$Cd_o = \left[\frac{k}{\ln\left(\frac{z}{z_{och}}\right)}\right]^2,$$ (56-2)

$$\beta' = 0.13 C d_o^{1/2} / k .$$ (56-3)

where Cd_o is the drag coefficient in the absence of wave effect, β' an empirical constant, c_p the phase speed at the peak frequency, c_p/u_* the wave age, and a_o another empirical constant. Eq. (56-1) shows the effect of wave age or swell on the drag coefficient. In Eq. (56-1), the drag coefficient, Cd, is expressed in terms of the drag coefficient Cd_o and the relative wave age, $c_p/u_* - a_o$, that is the wave age relative to the value of an empirical constant a_o. The drag coefficient Cd_o, in Eq. (56-2) is a result of the logarithmic wind profile under the neutral condition and z_{och} is the corresponding surface roughness. According to Large and Pond (1981), the expected average value of $Cd_o = 1.3 \times 10^{-3}$ and thus from Eq. (56-3), we obtain the value of $\beta' = 0.012$. A value of wave age, c_p/u_*, near 25 is associated with minimal wind-wave coupling influence (see Davidson, 1974). Therefore, we set the value of $a_o = 25$, which implies that when the wave age is equal to 25, $c_p/u_* = a_o$, the drag coefficient Cd is equal to Cd_o which is in the absence of swell effect. Theoretical justification of Eq. (56-1) is given by Brutsaert (1973). Eq. (56-1) indicates that when the wave age is greater than 25, i.e., $c_p/u_* > a_o$, the drag coefficient C_d decreases with increasing wave age; this condition corresponds to the older sea. When the wave age is less than 25, i.e., $c_p/u_* < a_o$, the drag coefficient C_d increases with decreasing wave age; this condition corresponds to the younger sea or developing waves.

By using Eq. (52) and from Eq. (56-1), we obtain the following equation for the surface roughness in a more general form:

$$z_o = \frac{\alpha}{g}(u_*^2 + \gamma' w_*^2) + 0.11\frac{v}{u_*}$$ (57-1)

where

$$\alpha = \frac{\alpha_{ch}}{\beta^* (\frac{c_p}{u_*} - \alpha_o)} ,$$ (57-2)

$$\approx \frac{\alpha_{ch}}{1 + \beta^* \left(\frac{c_p}{u_*} - \alpha_o\right)} , \text{ for } \beta^* \left(\frac{c_p}{u_*} - \alpha_o\right) < 1$$ (57-3)

Where β^* is an empirical constant and is equal to $\beta^* = k\beta'/cd_o^{1/2} = 0.13$. Eq. (57-2) implies that the modified Charnock parameter, a, or the non-dimensional roughness, and hence the surface roughness length, decrease with increasing wave age for $c_p/u_* > a_o$. Therefore, the singularity of the surface roughness is removed when the turbulent shear stress approaches zero, as indicated in Eq. (46) or Eq. (57-1) (if the viscous term is neglected). The non-dimensional roughness in Eq. (57-3) is for the value of $|\beta(c_p/u_* - a)| < 1$.

For young or developing waves, Toba et al. (1990) proposed the following formula for the non-dimensional surface roughness

$$\frac{gz_o}{u_*^2} = \gamma^* \frac{c_p}{u_*},$$

(58)

where γ^* is an empirical constant and is equal to 0.025. For young and growing waves, the non-dimensional surface roughness is also dependent on the wave age, c_p/u_*, in the form as indicated in Eq. (58), and furthermore, the non-dimensional roughness length is also related to the Charnock parameter, i.e., $gz_o/u_*^2 = a$; therefore, from Eq. (57-2) and Eq. (58), the non-dimensional roughness length can be expressed in the following functional form

$$\frac{gz_o}{u_*^2} = \frac{q'.\left(\frac{c_p}{u_*}\right)}{e^{\beta^*\left(\frac{c_p}{u_*}-\alpha_o\right)}}.$$

(59)

Where q' is an empirical constant that can be obtained from the experimental data. We obtain the value of $q' = 0.001$ $(= \gamma^* e^{-24\beta^*})$ by setting the wave age to $c_p/u_* = 1$ and fitting the curve through the cluster of the observational data (see Fig. 4.).

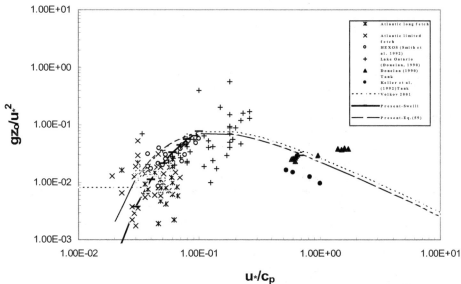

Fig. 4. Dependency of the non-dimensional roughness gz_o/u_*^2 on the inverse wave age u_*/c_p. Data are the collection of Donelan et al. (1993). Dotted line is from Volkov (2001), thin line from Eq. (59), and solid line in the presence of swell Eq. (57-2).

The semi-empirical formulas of the surface roughness suggested by Volkov (2001) are expressed as follows

$$\frac{gz_o}{u_*^2} = 0.03\left(\frac{c_p}{u_*}\right)^{-0.14\left(\frac{c_p}{u_*}\right)} e , \text{ for } 0.35 < \frac{c_p}{u_*} < 35 \qquad (60\text{-}1)$$

$$= 0.008 . \qquad \text{for } \frac{c_p}{u_*} > 35 \qquad (60\text{-}2)$$

In Eq. (60-2), it is assumed that the non-dimensional roughness is a constant value of gz_o/u_*^2 = 0.008 for the wave age $c_p/u_* > 35$, which implies that even under the conditions of swell, the non-dimensional roughness is a constant value of 0.008. Eq. (60-1) is the non-dimensional roughness in the absence of swell. The comparison of Eq. (59) and Eq. (60-1) with experimental data is given in Fig. 4; the experimental data were collected by Donelan (1993). The two datasets by Davidson (1974) and by Donelan et al. (1993) were obtained under the influence of heavy swell. The Atlantic Ocean data are dominated by swell (see Fig. 4 and Donelan et at., 1993). Eq. (59) was derived from the wave-dependent formulation of Davidson. The theoretical justification was also provided by Brusaert (1973).

As can be seen from Fig. 4, the result obtained from Eq. (59) is remarkably close to the roughness formula (Eq. 60-1) proposed by Volkov (2001), considering the different approaches used to derive the surface roughness lengths; the two curves are almost identical to each other. However, under the conditions of strong swell, the two curves diverge significanly. It appears that Eq. (57-2) has the better fit to the experimental data. The data in Fig. 4 show that under the influence of swell, the surface roughness is much less than the value of 0.008 proposed by Volkov (2001). That is, under the influence of heavy swell, Eq. (57-2) shows that the sea surface is very smooth, much smoother than suggested by Volkov (2001), i.e., there exits a "super" smooth sea surface.

Equations (35), (38), (39), (46), (49), (54), (56), (57), and (59) are the main results of this paper. In theory, the total stress is controlled mainly by the short to moderate gravity waves, the large eddies associated with the convective activity, the wave-induced stress, or the swell. For moderate wind speeds, the Charnock relation is a good approximation for the estimate of the surface roughness length. Under the conditions of swell, part of momentum flux may be transferred from the ocean to the atmosphere. The swell also has influence on the surface roughness and the total stress. Eq. (57) and Eq. (59) imply that the wave-induced stress is to modulate the Charnock parameter. For young and developing waves, the surface roughness increases, while for old sea with large wave age, the wave damping occurs. Abdella and D'Alessio (2003) conducted numerical experiments using the second-order turbulence closure scheme and the convection–induced stress in the Charnock relation; their results of sea surface temperature and heat content obtained from the model simulations are in good agreement with the observations. The equations (57-2) and (59), the non-dimensional surface roughness under the influence of swell, are for the first time theoretically derived in this study.

Concerning the practical applications of the proposed formulation for the surface roughness, it has been demonstrated that the approaches similar to the proposed approaches have already been used for practical applications, for examples, Janssen's formulation used in WAM model (the European Centre for Medium-Range Weather Forecasts (ECMWF); Hasselmann et al., 1988) and numerical simulations conducted by Abdella and D'Alessio (2003) under the light wind conditions. These results are in good agreement with the observational data.

7. Summary and conclusions

The Charnock wind stress formula is well known and has been widely used for the study of the air-sea interaction. In his formula, Charnock (1955) considered only forced convection and did not consider two other important physical processes that affect the surface roughness: free convection and swell. A parameterization of surface roughness length for the air-sea interaction in free convection has been suggested by Abdella and D'Alessio (2003). In the free convection limit, as the horizontal mean wind speed and the friction velocity vanish, the well-known Charnock wind stress formulation is inadequate and the traditional Monin-Obukhov similarity theory breaks down. According to Deardorff (1970), under free convection conditions, the convective velocity is an important scaling parameter; in the parameterization of the surface roughness, the velocity or the relevant similarity variables should be scaled by the convective velocity rather than the friction velocity. In free convection, the large-scale convective eddies in connection with the convective circulations, which are expected to reach near the surface, would induce the "virtual" surface wind stress, the so-called convection-induced stress, that, in turn, contributes to the increase of surface roughness.

In this study, two alternative approaches were used to derive the surface roughness length. In the first approach, we derived a new parameterization scheme of the sea surface roughness for the conditions of the forced and free convection based on the Prandtl mixing length theory and the standard deviation of the vertical turbulent velocity. In the derivation of this new parameterization scheme, first we introduced a new velocity scale, which squared value is equal to the "total stress" and is a combination of the friction velocity and the convective velocity. We then showed that, by replacing the friction velocity in the Charnock formula with the new velocity scale or the effective friction velocity, the Charnock formula can be extended to include the conditions of free convection, that is, to show that the surface roughness length depends on a new velocity scale, which is a combination of the forced convection and free convection. The second approach is based on the standard deviations of the horizontal velocity components to derive the relationship between the friction velocity and the convective velocity, which can be used to estimate the surface roughness length. This approach is consistent with the concept of gustiness proposed by Businger (1973) and others (see Schumann, 1988 and Sykes et al. , 1993). Friction velocity and convective velocity are also used as scaling parameters. Wind stress is in the horizontal direction that represents friction velocity; therefore, in a sense friction velocity (stress) and convective velocity are perpendicular to each other.

The dependence of the roughness length on the sea state and the effect of the wave-induced stress or the swell on the roughness length are also investigated in this paper. The influence of the wave-induced stress or the swell on the surface roughness is taken into account by modifying the Charnock parameter that depends on the wave age. For moderate wind speeds, the Charnock relation is a good approximation for the estimate of the surface roughness length. Abdella and D'Alessio (2003) performed the numerical simulations including the convective velocity; they showed that their results of sea surface temperature and heat content are in good agreement with the observations. For young and developing waves, the wave-induced stress enhances the surface roughness; the surface roughness in the absence of swell derived in this paper is remarkably close, almost identical to the wind stress formula proposed by Volkov (2001). However, in this paper, the proposed formula for the aerodynamic roughness in the presence of heavy swell, deviated significantly from

Volkov's formula, is for the first time theoretically derived and substantiated by experimental data; it shows that the surface roughness depends on the relative wave age. The results are in better agreement with experimental data. The results show that under the influence of heavy swell, there exists a very smooth sea surface, a "super" smooth sea surface, much smoother than that, a constant value, as suggested by Volkov (2001). Therefore, in this study, we have extended the Charnock wind stress relation to the forced and free convection in the presence of swell. An advanced wind-wave and atmospheric boundary layer measurement system has been deployed to an offshore oil platform in the Gulf of Mexico (Huang et al., 2011). The data collected from this wind-wave measurement system may shed additional light on this subject and further elucidate the behavior of air flow over ocean waves and the physical processes of the air-sea interaction.

8. Acknowledgments

The author appreciates the support of the U.S. Department of the Interior, Bureau of Ocean Energy Management, Regulation and Enforcement, Gulf of Mexico OCS Region, during the preparation of this manuscript. The opinions expressed by the author are his own and do not necessarily reflect the opinion or policy of the U.S. Government. The author also is indebted to the editor and anonymous reviewers for suggestions to improve the quality of this paper.

9. References

Abdella, K. and S. J. D. D'Alessio. (2003). A parameterization of the roughness length for the air-sea interaction in free convection. *Envir. Fluid Mechanics, 3,* 55-77.

Brusaert, W. (1973). Similarity functions for turbulence in neutral air above swell. *J. Phys. Oceanogr,* 4, 479-482.

Businger, J. A. (1973). A note on free convection. *Boundary-Layer Meteorol.,* 4, 323–326.

Charnock, H. (1955). Wind stress on a water surface. *Quart. J. Roy. Meteorol.* Soc., 81, 639-640.

Davidson, K. L. (1974). Observational results on the influence of stability and wind-wave coupling on momentum transfer and turbulent fluctuations over ocean waves. *Boundary- Layer Meteorol.,* 6, 305–331.

Deardorff, J. W. (1970). Convective velocity and temperature scales for the unstable boundary layer and for Rayleigh convection. *J. Atmos. Sci.,* 27, 1211-1213.

Donelan, M. A., F. W. Dobson, S. D. Smith and R. J. Anderson. (1993). On the dependence of sea surface roughness on wave development. *J. Phys. Oceanogr.,* 23, 2143-2149.

Dyer, A. J. (1974). A review of flux-profile relationships. *Boundary-Layer Meteorol.,* 7, 363-372.

Fairall, C, W., E. F. Bradley, D. P. Rogers, J. B. Edson and G. S. Young. (1996). Bulk parameterization of air-sea fluxes for Tropical Ocean-Global Atmosphere Coupled-Ocean Atmosphere Response Experiment. *J. Geophys. Res.,* 101(C2), 3747-3764.

Garratt, J. R. (1977). Review of drag coefficients over oceans and continents. *Mon Wea. Rev., 105,* 915-929.

Garratt, J. R. (1992). *The atmospheric Boundary Layer.* Cambridge University Press, Cambridge, 316 pp.

Hogstrom, U. (1988). Non-dimensional wind and temperature profiles in the atmospheric surface layer: A re-evaluation. *Boundary-Layer Meteorol.,* 42, 55-78.

Huang, C.H. (1979). A theory of dispersion in turbulence shear flow. *Atmospheric Envir.*, 13, 453-463.

Huang, C. H. (2009). Parameterization of the roughness length over the sea in forced and free convection. *Environmental Fluid Mech.*, 2009, 9:359-366.

Huang, C. H., C. MacDonald, Randolph "Stick" Ware, A. Ray, C. A. Knoderer, J. Hare, W. Gibson, L. Bariteau, P. Roberts, C.W. Fairall, S. Pezoa. (2011). Thermodynamic Profiling and Temporal Evaluation of Mixing Height during the Cold Air Outbreak in the Gulf of Mexico. *Thermodynamic Profiling Technologies Workshop (poster session).* April 12-14, 2011 held at Boulder, Colorado.

Janssen, P. A. E. M. (1989). Wave-induced stress and the drag of air flow over sea waves. *J. Phys. Oceanogr.*, 19, 745-754.

Janssen, P. A. E. M. (1991). Quasi-linear theory of wind-wave generation applied to wave forecasting. *J. Phys. Oceanogr.*, 21, 1631-1642.

Johnson, H. K., J. Hojstrup, H. J. Vested and S. E. Larsen. (1998). On the dependence of sea surface roughness on wind waves. *J. Phys. Oceanogr.*, 28, 1702-1716.

Large, W. G. and S. Pond. (1981). Open ocean momentum flux measurements in moderate to strong winds. *J. Phys. Oceanogr.*, 11, 324-336.

Makin V. K. (2008). On the possible impact of a following-swell on the atmospheric boundary layer. *Boundary-Layer Meteorol.* 129:469–78.

Makin V. K. & Kudryavtsev, V. N. (1999). Coupled sea surface-atmosphere model. Part 1. Wind over waves coupling. *J. Geophys. Res.*, 104(C4), 7613--7623.

Miles, J. W. (1957). On the generation of surface waves by shear flow. *J. Fluid Mech.*, 3, 185-204.

Monin, A. S. and A. M. Obukhov. (1954). Basic turbulence mixing laws in the atmospheric surface layer. *Trudy Geofiz. Inst. AN SSSR.*, No. 24 (151), 163-187.

Monin, A .S. and A. M. Yaglom. (1971). *Statistical fluid mechanics; mechanics of turbulence*, MIT Press, Cambridge, MA. 769 pp.

Nikuradse, J. (1933). Stromungs gesetze in rauhen Rohren. *Forschungsheft*, No. 361.

Panofsky, H. A., H. Tennekes, D. H. Lenschow and J. C. Wyngaard. (1977). The characteristics of turbulent velocity components in the surface layer under convective conditions. *Boundary-Layer Meteorol.*, 11, 355-361.

Paulson, C.A. (1970): The mathematical representation of wind speed and temperature profiles in the unstable atmospheric surface layer, *J. Appl. Meteorol.*, 9, 857-861.

Obukhov, A. M. (1946). Turbulence in an atmosphere with a non-homogenous temperature. *Trudy Inst. Theor. Geophy.*, USSR, 1:95-115.

Phillips, O. M. (1966). *The Dynamics of the Upper Ocean.* Cambridge Univ. Press, Cambridge, U.K., 261 pp.

Phillips, O. M. (1977). *The Dynamics of the Upper Ocean*, 2nd ed. Cambridge Univ. Press, Cambridge, U.K., 336 pp.

Schumann, U. (1988). Minimum friction velocity and heat transfer in the rough surface layer of a convective boundary layer. *Boundary-Layer Meteorol.*, 44, 311-326.

Smith, S. D. (1980). Wind stress and heat flux over the ocean in gale force winds. *J. Phys. Oceanogr.*, 10, 709-726.

Smith, S. D. (1988). Coefficients for sea surface wind stress, heat flux, and wind profiles as a function of wind speed and temperature. *J. Geophys. Res.*, 93, 15467-15472.

Smith, S. D., R. J. Anderson, W. A. Oost, C. Kraan, N. Maat, J. DeCosmo, K. B. Katsaros, K. L. Davison, K. Bumke, L. Hasse and H. M. Chadwick. (1992). Sea surface wind stress and drag coefficients: the HEXOS results. *Boundary-Layer Meteorol.*, 60, 109-142.

Stull, B. R. (1988). *An Introduction to Boundary Layer Meteorology.* Kluwer Academic Publishers. 666 p.

Sykes, R. I., D. S. Henn and W. S. Lewellen. (1993). Surface-layer description under free convection conditions. *Quart. J. Roy. Meteorol.* Soc., 119, 409–421.

Troen, I. and L. Mahrt. (1986). A simple model of the atmospheric boundary layer: sensitivity to surface evaporation. *Boundary-Layer Meteorol.*, 37, 129-148.

Toba, Y., N. Iida, H. Kawamura, N. Ebuchi and I. S. F. Jones. (1990). Wave dependence of sea-surface stress. *J. Phy. Oceanogr.*, 20, 705-721.

Volkov, Y. (2001). The dependence on wave age. p. 206-217. *In Wind Stress over the Ocean*, ed. by I. S. F. Jones and Y. Toba. Cambridge Univ. Press, New York, 307 pp.

Wyngaard, J. C., and O. R. Coté. (1974). The evolution of a convective planetary boundary layer-A higher-order-closure model study. *Boundary.-Layer Meteorol.*, 7, 289–308.

WAMDI group: Hasselmann, S., K. Hasselmann, E. Bauer, P.A.E.M Janssen, G.J. Komen, L. Bertotti, P. Lionello, A. Guillaume, V.C. Cardone, J.A. Greenwood, M. Reistad, L. Zambresky and J. A. Ewing. (1988). The WAM model-a third generation ocean wave prediction model. *J. Phys. Oceanogr.*, 18, 1775- 1810.

Wilson D. K. (2001). An alternative function for the wind and temperature gradients in unstable surface layers. *Boundary-Layer Meteorol.*, 99, 151-158.

7

Radial Basis Functions Methods for Solving Radionuclide Migration, Phase Change and Wood Charring Problems

Leopold Vrankar[1], Franc Runovc[1] and Goran Turk[2]
[1]Faculty of Natural Sciences and Engineering,
[2]Faculty of Civil and Geodetic Engineering,
University of Ljubljana
Slovenia

1. Introduction

Modelling the flow through porous media has a great importance for solving the problems of disposal of radioactive waste. When modelling the flow of contaminated material through the geosphere, it is important to consider all internal processes (e.g. advection, dispersion, retardation) within the geosphere, and external processes associated with the near-field and the biosphere. The general reliability and accuracy of transport modelling depend predominantly on input data such as hydraulic conductivity, water velocity on the boundary, radioactive inventory, hydrodynamic dispersion. The output data are concentration, pressure, etc. The most important input data are obtained from field measurement, which are not available for all regions of interest. In such cases, geostatistical science offers a variety of spatial estimation procedures.

A vast variety of important physical processes involving heat conduction and materials undergoing a change of phase may be approached as Stefan problems. One of these processes is the heat transfer involving phase changes caused by solidification or melting, which are important in many industrial applications such as the drilling of high ice-content soil, the storage of thermal energy and the safety studies of nuclear reactors. Due to their wide range of applications, the phase change problems have drawn considerable attention of specialists in different fields of science and engineering.

Another problem which seems to be completely different but is in mathematical terms very similar to solidification or melting is charring of wood. After wood is exposed to fire it undergoes thermal degradation. The pyrolysis gases undergo flaming combustion as they leave the charred wood surface. The pyrolysis, charring, and combustion of wood have been presented by (Fredlund, 1993) who performed experiments and numerical analyses.

For all physical processes mentioned above, the motion of fluids, phase changes, and pyrolysis processes are governed by a set of partial differential equations (PDES). These governing equations are based upon the fundamental conservation laws. The mass, momentum and energy are conserved in any fluid motions. In most cases, the governing equations are too complex to be solved analytically. Therefore the numerical methods have been extensively used to find an approximate solution of the equations.

Traditionally, the most popular methods have been the finite element methods (FEM), the finite difference methods (FDM), and boundary element method (BEM). In recent years, the radial basis functions (RBF) methods have emerged as a novel computing method in scientific computing community. All the conventional methods mentioned above can be considered as mesh-based methods characterized by their reliance on a computation mesh with certain relationship between the nodes. In spite of their great success in solving scientific and engineering problems over the past five decades, these conventional numerical methods still have some drawbacks that impair their computational efficiency and even limit their applicability to more practical problems, particularly in three-dimensional space. The RBF methods could overcome some of these drawbacks by constructing the approximations entirely in terms of a set of points. The methods can be extended to multi-dimensional problems without significant effort. Over the last 30 years, many researchers have shown a great deal of interest in RBFs. It was used for modelling radionuclide migration (Vrankar et al., 2004a), (Vrankar et al., 2005), structural topology optimization (Wang S & Wang MY, 2006) and many other applications (Ling, 2003).

The chapter is organized as follows. In the first section a short description of geostatistics is given. The sections on radial basis functions follow. The efficiency of the presented method is explained on three very different cases. The first example presents the usage of RBFs in geostatistical analysis for modelling of the radionuclide migration. The advection-dispersion equation is used in either Eulerian or Lagrangian form.

The second case presents a phase change problem. The moving interface is captured by the level set method at all time with the zero contour of a smooth function known as the level set function. A new approach is used to solve a convective transport equation for advancing the level set function in time. This new approach is based on the asymmetric meshless collocation method and the adaptive combination of RBFs method and Greedy algorithm for trial subspaces selection.

The last case explores the char formation in the wood as a function of surrounding temperature. The problem is solved numerically by the radial base function (RBF) methods. The results are tested on the one-dimensional case in the standard fire conditions, and the same model is used to analyse a two-dimensional behaviour of timber beam exposed to fire from three sides.

At the end of the chapter some conclusions are given.

2. Geostatistics

The term geostatistics is employed here as a generic term, meaning the application of the theory of random fields in the earth sciences (Kitanidis & VoMvoris, 1983). The parameters are distributed in space and can thus be called regionalized variables. The parameters of a given geologic formation can conveniently be represented as realisations of random variables which form random fields.

Stochastic simulation is a widely accepted tool in various areas of geostatistics. Simulations are termed globally accurate through the reproduction of one-, two-, or multiple-point statistics representative of the area under study.

The most convenient method for simulation of random fields is sequential Gaussian simulation (Deutsch & Journel, 1998) because all successive conditional distributions from which simulated values are drawn are Gaussian with parameters determined by the solution of a simple kriging system.

Sequential Gaussian simulation procedure:

1. First, use a sequential Gaussian simulation to transform the data into a normal distribution.

2. Perform variogram modelling on the data. Select one grid node at random, then krige the value at that location. This also gives the kriged variance.

3. Draw a random number from a normal distribution that has a variance equivalent to the kriged variance and a mean equivalent to the kriged value. This number is the simulated number for that grid node.

4. Select another grid node at random and repeat. For the kriging, include all the previously simulated nodes to preserve the spatial variability as modelled in the variogram.

5. When all nodes have been simulated, back transform to the original distribution. This gives us first realization using a different random number sequence to generate multiple realizations of the map.

Kriging (D. G. Krige, a South African mining engineer) is a collection of generalized linear regression techniques for minimizing an estimation variance defined from a prior model for a covariance (semivariogram) (Olea, 1991). One of the basic statistical measures of geostatistics is the semivariance, which is used to express the rate of chance of a regionalized variable along a specific orientation (or a measure of degree of spatial dependence between samples along a specific support) (Davis, 1986). If we calculate the semivariances for different values of h, we can plot the results in the form of a semivariogram:

$$\gamma(h) = 0.5 \times Var\left[Z(\mathbf{x} + \mathbf{h}) - Z(\mathbf{x})\right], \tag{1}$$

where $Z(\mathbf{x})$ and h are the realization of random process in the space and norm of vector $h = (\|\mathbf{h}\|)$. Since the semivariogram is a function of distance, the weights change according to the geographic arrangement of the samples. Kriging can be used to make contour maps, but unlike conventional contouring algorithms, it has certain statistically optimal properties.

3. Meshless methods

So far, research on the numerical method has focused on the idea of using a meshless methodology for the numerical solution of PDEs. One of the common characteristics of all mesh-free methods is their ability to construct functional approximation or interpolation entirely from information at a set of scattered nodes, among which there is no relationship or connectivity needed:

Three approaches to meshless methods have been successfully proposed. The first one is based on the finite element method and employs Petrov-Galerkin weak formulation (Atluri & Shen, 2002). The second one is of boundary element type (Li & Aluru, 2002). The third approach employs the RBFs. The base of this approach is its employment of different interpolating functions to approximate solutions of differential equations. Kansa (1990a) introduced multiquadric functions to solve hyperbolic, parabolic and elliptic differential equations with collocation methods. One of the most powerful RBF methods is based on multiquadric basis functions (MQ), first used by (Hardy, 1971).

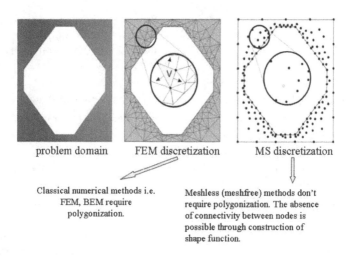

| problem domain | FEM discretization | MS discretization |

Classical numerical methods i.e. FEM, BEM require polygonization.

Meshless (meshfree) methods don't require polygonization. The absence of connectivity between nodes is possible through construction of shape function.

Fig. 1. Graphical presentation of Classical vs. Meshless methods

A radial basis function is the function

$$\varphi_j(\mathbf{x}) := \varphi(||\mathbf{x} - \mathbf{x}_j||),$$

which depends only on the distance between $\mathbf{x} \in \mathbb{R}^d$ and a fixed point $\mathbf{x}_j \in \mathbb{R}^d$. Here, φ_j is continuous and bounded on any bounded sub-domain $\Omega \subseteq \mathbb{R}^d$ whereas $\varphi : \mathbb{R}^d \to \mathbb{R}$. Let $r \geq 0$ denote the Euclidean distance between any pair of points in the domain Ω. The commonly used radial basis functions are: linear $(\varphi(r) = r)$, cubic $(\varphi(r) = r^3)$, thin-plate spline $(\varphi(r) = r^2 \log r)$ and Gaussian $(\varphi(r) = e^{-\alpha r^2})$. The most popular globally supported C^∞ RBFs are MQ $(\varphi(r) = (1 + (r/c)^2)^\beta)$, $\beta = 1/2$), (Wang & Liu, 2002). It is also important to mention the local version of the RBF collocation which does not produce the ill conditioning and is less sensitive to the shape parameter selection. The local version of the RBF collocation is based on compactly supported RBF spline: $(\varphi(r) = (1 - r)^m_+ p(r))$ where $p(r)$ is a polynomial of the Wendland (Wendland, 1995) compactly supported (CS-RBF) spline.

The parameter $c > 0$ is a shape parameter controlling the fitting of a smoothing surface to the data. It has a significant influence on the accuracy of the solution. For this reason, in almost all previous researches the shape parameter must be adjusted with the number of centers in order to produce equation systems that are sufficiently well conditioned to be solved with the standard finite precision arithmetic. The optimal choice of the constant shape parameter is still an open question, and it is most often selected by the trial and error approach.

To overcome the problems of ill-conditioned matrix many efforts have been made to find new computational methods being capable of avoiding the ill-conditioning problems using linear solvers. In the literature, the following methods are reported:

- (a) Using variable shape parameters (Kansa, 1990b),

- (b) preconditioning the coefficient matrix, see Ling and Kansa (Ling & Kansa, 2004),

- (c) using domain decomposition methods in overlapping or non-overlapping schemes that decompose a very large ill-conditioned problem into many subproblems with better conditioning (Kansa, 1990a),

- (d) optimizing the center locations by the Greedy algorithm.

- (e) using an improved numerical solver based on affine space decomposition (Ling & Hon, 2005),

- (f) using complex MQ shape parameters (Fornberg & Wright, 2004), etc.

Combination of the above can be used as optimized solution procedure and accuracy.

3.1 Asymmetric meshless collocation methods for stationary problems

We briefly review the RBF asymmetric collocation scheme. We consider a PDE in the general form of

$$Lu(\mathbf{x}) = f(\mathbf{x}), \quad \text{in } \Omega \subset \mathbb{R}^d, \tag{2}$$

$$Bu(\mathbf{x}) = g(\mathbf{x}), \quad \text{on } \partial\Omega, \tag{3}$$

where u is the unknown solution, d denotes the dimension, $\partial\Omega$ is the boundary of the domain Ω, L is the differential operator on the interior, and B is the operator that specifies the boundary conditions of the Dirichlet, Neumann or mixed type. Both, f and g, are given functions with sufficient smoothness mapping $\mathbb{R}^d \mapsto \mathbb{R}$. Using the Kansa's asymmetric multiquadric collocation method, the unknown PDE solution u is approximated by RBFs in the form:

$$u \approx U(\mathbf{x}) = \sum_{j=1}^{N} \alpha_j \varphi_j(\mathbf{x}) + \sum_{l=1}^{Q} \gamma_l v_l(\mathbf{x}), \tag{4}$$

where φ is any type of radial basis function, and $v_1, ..., v_M \in \Pi_q^d$, are polynomials of degree m or less, $Q := \begin{pmatrix} q - 1 + d \\ d \end{pmatrix}$, see (Iske, 1994). Let $(\mathbf{x}_j)_{j=1}^{N}$ be the N collocation points in $\Omega \cup \partial\Omega$. We assume the collocation points are arranged in such a way that the first N_I points are in Ω, whereas the last N_B points are on $\partial\Omega$. To evaluate or approximate the $N + Q$ unknown coefficients, at least, $N + Q$ linearly independent equations are needed. Ensuring that $U(\mathbf{x})$ satisfies equations (2) and (3) at the collocation points results in a good approximation of the solution u. The first N equations are given by

$$\sum_{j=1}^{N} \alpha_j L\varphi_j(\mathbf{x}_i) + \sum_{l=1}^{Q} \gamma_l Lv_l(\mathbf{x}) = f(\mathbf{x}_i), \quad \text{for } i = 1, ..., N_I, \tag{5}$$

and the others are given as

$$\sum_{j=1}^{N} \alpha_j B\varphi_j(\mathbf{x}_i) + \sum_{l=1}^{Q} \gamma_l Bv_l(\mathbf{x}) = g(\mathbf{x}_i), \quad \text{for } i = N_I + 1, ..., N_I + N_B. \tag{6}$$

The last Q equations could be obtained by imposing additional conditions on $p(\cdot)$:

$$\sum_{j=1}^{N} \alpha_j v_k(\mathbf{x}_j) = 0, \quad k = 1, ..., Q. \tag{7}$$

The above procedure leads to the system of equations

$$\begin{bmatrix} W_L & p_L \\ W_B & p_B \\ p^T & 0 \end{bmatrix} \begin{bmatrix} \alpha \\ \gamma \end{bmatrix} = \begin{bmatrix} f \\ g \\ 0 \end{bmatrix}, \tag{8}$$

where

$$W_L = L\varphi_j(\mathbf{x}_i), \quad \mathbf{x}_i \in X_I, \tag{9}$$
$$p_L = Lv_l(\mathbf{x}_i), \quad \mathbf{x}_i \in X_I, \tag{10}$$
$$W_B = B\varphi_j(\mathbf{x}_i), \quad \mathbf{x}_i \in X_B, \tag{11}$$
$$p_B = Bv_l(\mathbf{x}_i), \quad \mathbf{x}_i \in X_B, \tag{12}$$
$$p = v_k(\mathbf{x}_i), \quad k = 1, ..., Q. \tag{13}$$

The RBF method for solving PDEs is straightforward and easy to implement. Since the RBF is differentiable, spatial derivatives are computed by simply differentiating the MQ-RBFs; time derivatives are computed by differentiating the time dependent expansions coefficients. The spatial and temporal partial derivatives result in either a linear or nonlinear set of ordinary differential equations.

4. Modelling of the radionuclide migration

The central issue in modelling is on the one hand consistency between conceptual and mathematical models and, on the other hand between conceptual models and scenarios. A conceptual model is a qualitative description of the system functioning in a form which corresponds to mathematical representation. Each scenario is a set of features, processes and events which has to be considered together to assess the impact of the disposal in the future. Groundwater models are presented by motion and continuity equations. The majority of the codes currently used or under development are based on the advective-dispersive equation (Bear, 1972) with various physical phenomena added. According to this equation, mass transport is controlled by two mechanisms: advection and dispersion. Advection governs the movement of the solute, linked to the fluid, with the water velocity. Water velocity can be assessed through Darcy's law. Dispersion represents mixing of diffusion and random variations from the mean stream.

The simulation area will be 2D rectangular with the Neumann and Dirichlet boundary conditions. The Neumann boundary conditions represent flow while Dirichlet boundary conditions represent constant pressure and concentration.

4.1 Eulerian form of the advection-dispersion equation

The first step of radionuclide transport modelling is to obtain the Darcy velocity, here denoted by $\mathbf{V}$. The continuity equation (e.g. for fluid phase) can be written

$$\frac{\partial(\epsilon \varrho_f)}{\partial t} + \nabla \cdot (\varrho_f \mathbf{V}) = 0, \tag{14}$$

where ϱ_f is the density of fluid. If the porosity ϵ is constant, equation (14) is simplified to

$$\epsilon \frac{\partial(\varrho_f)}{\partial t} + \nabla \cdot (\varrho_f \mathbf{V}) = 0. \tag{15}$$

With assumed constant density, Equation (15) simplifies to

$$\nabla \cdot \mathbf{V} = 0. \tag{16}$$

In saturated porous media the fluid flow is described by the Darcy equation

$$\nabla p = -\frac{\mu}{\mathbf{K}} \mathbf{V} = 0, \tag{17}$$

where ∇p is the pressure gradient, $\mathbf{K}$ is hydraulic conductivity tensor, and μ is the dynamic viscosity. If we assume that the hydraulic conductivity is constant in each point of the considered area, then the combination of equations (16) and (17) gives a type of Laplace equation. In cases of homogeneous or heterogeneous and anisotropic porous media and incompressible fluid, the calculation of the velocities in principal directions which were determined from the pressure of the fluid obtained from the Laplace differential equation is presented in (Vrankar et al., 2004a), (Vrankar et al., 2005). The Lagrangian form of the advection-dispersion equation is presented in (Vrankar et al., 2004b)

The velocities obtained from Laplace equation are used in the advection-dispersion equation. The advection-dispersion equation for transport through the saturated porous media zone with retardation and decay is:

$$R\frac{\partial u}{\partial t} = \left(\frac{D_x}{\omega} \frac{\partial^2 u}{\partial x^2} + \frac{D_y}{\omega} \frac{\partial^2 u}{\partial y^2} \right) - v_x \frac{\partial u}{\partial x} - R\lambda u, \quad (x,y) \in \Omega, \quad 0 \leq t,$$

$$u|_{(x,y)\in\partial\Omega} = g(x,y,t), \qquad\qquad\qquad 0 \leq t \tag{18}$$

$$u|_{t=0} = h(x,y), \qquad\qquad\qquad (x,y) \in \Omega,$$

where x is the groundwater flow axis, y is the transverse axis, u is the concentration of contaminant in the groundwater $[\text{Bqm}^{-3}]$, D_x and D_y are the components of dispersion tensor (two processes are incorporated: molecular diffusion and dispersion. Molecular diffusion is a thermo-chemical process, where mass is transported due to thermal or solutes gradients. On the other hand, dispersion is mechanical process, where spreading of the substance is caused due to the motion of the fluid. Details you can found in (Bear, 1972)) $[\text{m}^2\text{y}^{-1}]$ in the saturated zone, ω is porosity of the saturated zone $[-]$, v_x is Darcy velocity $[\text{my}^{-1}]$ at interior points, R is the retardation factor in the saturated zone $[-]$ and λ is the radioactive decay constant $[\text{y}^{-1}]$. In these cases $[\text{y}]$ means years.

For the parabolic problem, we consider the implicit scheme:

$$R\frac{u^{n+1} - u^n}{\delta t} = \left(\frac{D_x}{\omega} \frac{\partial^2 u^{n+1}}{\partial x^2} + \frac{D_y}{\omega} \frac{\partial^2 u^{n+1}}{\partial y^2} \right) - v_x \frac{\partial u^{n+1}}{\partial x} - R\lambda u^{n+1}, \tag{19}$$

where δt is the time step and u^n and u^{n+1} are the contaminant concentrations at the time t_n and t_{n+1}.

The approximate solution is expressed as :

$$u(x, y, t_{n+1}) = \sum_{j=1}^{N} \alpha_j^{n+1} \varphi_j(x, y), \tag{20}$$

where $\alpha_j^{n+1}, j = 1, ..., N$ are the unknown coefficients to be determined. $\varphi_j(x, y)$ is Hardy's multiquadrics function Hardy (1971):

$$\varphi_j(x, y) = \sqrt{(x - x_j)^2 + (y - y_j)^2 + c^2}, \tag{21}$$

where c is shape parameter.

By substituting (20) into (18), we have:

$$\sum_{j=1}^{N} \left(R\frac{\varphi_j}{\delta t} - \frac{D_x}{\omega}\frac{\partial^2 \varphi_j}{\partial x^2} - \frac{D_y}{\omega}\frac{\partial^2 \varphi_j}{\partial y^2} + v_x\frac{\partial \varphi_j}{\partial x} + R\lambda\varphi_j \right)\bigg|_{x_i, y_i} \alpha_j^{n+1} = R\frac{u^n(x_i, y_i)}{\delta t}, \tag{22}$$

$$i = 1, 2, ..., N_I.$$

$$\sum_{j=1}^{N} \varphi_j(x_i, y_i)\alpha_j^{n+1} = g(x_i, y_i, t_{n+1}), \qquad i = N_I + 1, N, \tag{23}$$

from which we can solve the $N \times N$ linear system of (22)–(23) for the unknown coefficients $\alpha_j^{n+1}, j = 1, ..., N$. Let $N = N_I + N_B$ be the number of collocation points, N_I is the number of interior points and N_B is the number of boundary points. Then (20) gives the approximate solution at any point in the domain Ω.

The traditional finite difference method (FDM) was also used for solving the Laplace and advection-dispersion equation. For the approximation of the first derivative second-order central difference or one-sided difference were used. But for the approximation of the second derivatives we used the second-order central difference.

4.2 Numerical examples

2D Solution

The simulation was performed on a rectangular area of 450 by 300 m. The source (initial condition) was Thorium $(Th - 230)$ with activity $1 \cdot 10^6 Bq$ and half life of 77000 years. The distribution of hydraulic conductivity (values between 66.00 and 83.41 $[\frac{m}{y}]$) for one specific simulation is shown in Fig. 2.

The groundwater flow field is presented for steady-state conditions. Except for the inflow (left side) and outflow (right side), all boundaries have a no-flow condition. The flow velocity was 1 m/y. At the outflow side, time-constant pressures at the boundaries were set. Longitudinal dispersivity D_x is 200 $[\frac{m^2}{y}]$ and transversal dispersivity D_y is 2 $[\frac{m^2}{y}]$. For the porosity ϵ we used values between 0.25 and 0.26. The retardation constant R is 800. The distribution of the velocities calculated with FDM and MQ methods are presented in Fig. 3. The distribution of the average value of contaminant concentration after 100,000 years is also given in Fig. 3.

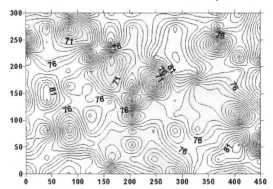

Fig. 2. Distribution of hydraulic conductivity

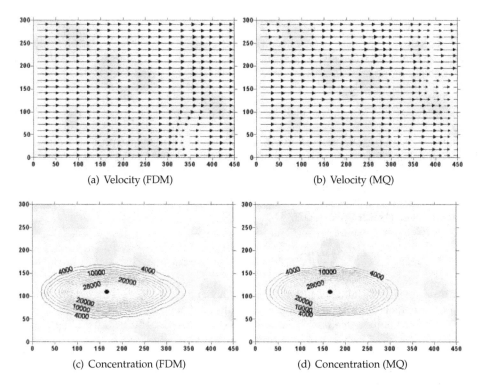

(a) Velocity (FDM)

(b) Velocity (MQ)

(c) Concentration (FDM)

(d) Concentration (MQ)

Fig. 3. Distribution of the velocities and avarage concentrations (symbol • shows the location of Thorium source)

5. Moving-boundary problems and Level Set Method(LSM)

The location of the solid-liquid interface for many phase change problems is not known *a priori* and must be determined during the course of the analysis. Mathematically, the motion of the interface is expressed implicitly in an equation for the conservation of thermal energy at the interface, the so-called Stefan conditions. These conditions causes the system to become non-linear, and therefore each problem is somewhat unique. Moving boundaries are also associated with time-dependent problems and the position of the boundary has to be determined as a function of time and space.

A level set method has become an efficient tool for tracking, modelling and simulating the motion of free boundaries in fluid mechanics, combustion, computer animation and image processing (Osher & Fedkiw, 2003), (Sethian, 1999). The objective of this part of the chapter is to present the combination of RBF approach and the level set method for solving two-dimensional moving-boundary problems.

The LSM is a numerical technique for tracking shapes and interfaces. The advantage of the LSM is that one can perform numerical computations involving curves and surfaces on fixed Cartesian grid. The LSM makes it very easy to follow shapes that change topology (e.g., when a shape splits in two or develops holes).

5.1 Level set function and equation

In two dimensions, the LSM represents a closed curve $\Gamma \in \mathbb{R}^2$ in the plane as the zero level set of a two-dimensional auxiliary function $\Phi(\mathbf{x}, t) : \mathbb{R}^2 \times \mathbb{R} \to \mathbb{R}$, where $\mathbf{x}$ is a position of the interface, t is a moment in time. Therefore, the closed curve is presented as:

$$\Gamma = \{(\mathbf{x}, t) | \ \Phi(\mathbf{x}, t) = 0\}. \tag{24}$$

The function Φ is also called a level set function and is assumed to take positive values inside the region delimited by the curve Γ and negative values elsewhere (Osher & Fedkiw, 2003), (Sethian, 1999). The level set function can be defined as a signed distance function to the interface. The moving interface is then captured at all time by locating the set of $\Gamma(t)$ for which Φ vanishes. The movement of the level set function can be described as the following Cauchy problem (Tsai & Osher, 2003):

$$\frac{\partial \Phi}{\partial t} + v^T \nabla \Phi = 0, \quad \Phi(\mathbf{x}, 0) = \Phi_0(\mathbf{x}), \tag{25}$$

where $\Phi_0(\mathbf{x})$ means the initial position of the interface and $v^T = [v_1, v_2]$ is the continuous field, which is a function of position $\mathbf{x}$. The above partial differential equation is often solved by using the finite difference method (FDM) on Cartesian grids (Osher & Fedkiw, 2003), (Sethian, 1999).

5.2 Level set equation construction with RBFs

The level set equation can be constructed in many ways. In our case, the Crank-Nicolson implicit scheme will be presented. The classical time stepping schemes are stabilized with the adaptive greedy algorithm, see (Vrankar et al., 2010).

We consider the Crank-Nicolson implicit scheme of (25):

$$\frac{\Phi^{n+1} - \Phi^n}{\Delta t} + \frac{1}{2}\left(v_1\frac{\partial\Phi^{n+1}}{\partial x} + v_2\frac{\partial\Phi^{n+1}}{\partial y} + v_1\frac{\partial\Phi^n}{\partial x} + v_2\frac{\partial\Phi^n}{\partial y}\right) = 0, \tag{26}$$

where $t_{n+1} = t_n + \Delta t$, Φ^{n+1} and Φ^n are the level set variable at time t_{n+1} and t_n. The approximate solution is expressed as:

$$\Phi(\mathbf{x}, t_{n+1}) = \sum_{j=1}^{N} \alpha_j^{n+1}\varphi_j(\mathbf{x}), \tag{27}$$

where α_j^{n+1}, $j = 1, ..., N$, are the unknown coefficients to be determined. By substituting equation (27) into (26), we obtain:

$$\sum_{j=1}^{N}\left(\frac{\varphi_j(\mathbf{x}_i)}{\Delta t} + \frac{1}{2}\left(v_1\frac{\partial\varphi_j(\mathbf{x}_i)}{\partial x} + v_2\frac{\partial\varphi_j(\mathbf{x}_i)}{\partial y}\right)\right)\alpha_j^{n+1} = \frac{\Phi^n(\mathbf{x}_j)}{\Delta t}$$

$$-\frac{1}{2}\left(v_1\frac{\partial\Phi^n(\mathbf{x}_j)}{\partial x} + v_2\frac{\partial\Phi^n(\mathbf{x}_j)}{\partial y}\right) \quad i = 1, ..., N. \tag{28}$$

5.3 Numerical examples

Oriented flow

The first example is translation of circular interface/bubble (the term bubble is connected to the study of multiphase flows (ETH, 2003). In the multi-fluid model, the motion of the specific dispersed phase elements (bubbles, drops) is not followed; rather, the dispersed phase elements which interact with the continuous phase flow field are observed. Therefore, computational modelling is assuming a greater role in the study of multiphase flows.) in oriented flow in Fig. 4. The circular interface of radius $r = 0.15$, initially centered at $(0.2, 0.7)$ and moved by the oriented flow in a cavity of size 1×1 with the velocity field (u, v) defined as follows:

$$u = -0.2(x + 0.5) \tag{29}$$
$$v = 0.2(y + 0.5). \tag{30}$$

For capturing the moving circular interface in Fig. 3, LMS and RBFs are used. The calculation is performed up to $t = 2$, in computational grid 20×20 (upper part of the Fig. 4) and 30×30 (lower part of the Fig. 4). The function Φ is evaluated on a 38×38 grid in order to find zero contours. We choose the following shape parameters $c = 0.033$, $c = 0.5$, and $c = 2$ to ensure smoothness of the circular interfaces over time and the initial time step is 0.01.

Shear flow

We next consider a circular interface (Fig. 5) of radius $r = 0.15$, initially centered at $(0.5, 0.7)$ and moved by a shear flow in a cavity of size 1×1 with the velocity field (u, v) defined as

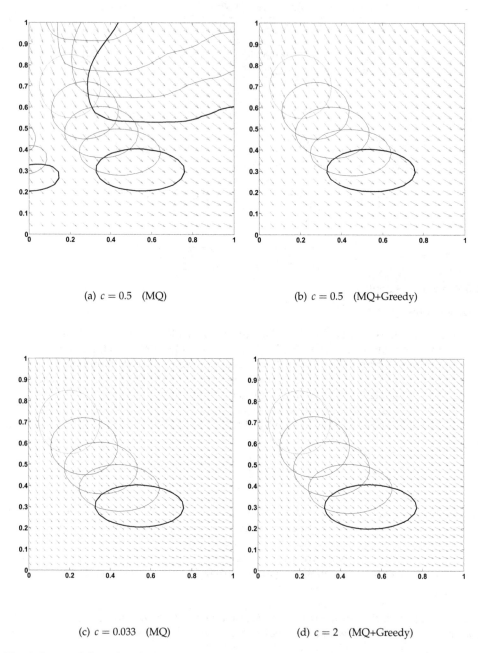

(a) $c = 0.5$ (MQ)   (b) $c = 0.5$ (MQ+Greedy)

(c) $c = 0.033$ (MQ)   (d) $c = 2$ (MQ+Greedy)

Fig. 4. Oriented flow distribution

follows:

$$u = \sin(\pi x)\cos(\pi y), \tag{31}$$

$$v = -\cos(\pi x)\sin(\pi y). \tag{32}$$

In such a velocity field, the circular interface is passively transported in the form of circulation and stretching. The LMS and RBFs are used to capture the moving interface with time, up to $t = 4$ with a time step $\triangle t = 0.01$, in computational grids 20×20 and 30×30. The function Φ is then evaluated on a 38×38 grid in order to find zero contours. We again choose the large shape parameter $c = 2$ to demonstrate stability over ill-conditioned linear systems.

In Fig. 5 , we can see that with or without the adaptive algorithm, the circular interfaces are more or less at the same location. However, the circular interface obtained with the adaptive algorithm is more skewed; without adaptive algorithm, a more circular bubble is obtained. We can see that some artifacts appear in the corner of the computational domain.

6. Wood charring problems

The model which explains a very complex phenomena of wood charring consists of the differential equation for heat transfer with corresponding boundary conditions, which prescribe the heat flow on the exposed boundaries of the cross-section. In our case, different types of boundary conditions were used, e.g. Dirichlet, Neumann, radiation term, etc. The char formation in the timber beam as a function of its temperature is taken into account by the model.

Since the analytical solution is seldom obtainable, the problem is solved numerically by the RBF methods (e. g. non-symmetric RBF collocation). Picard's or Newton's methods are used for the solution of the second order non-linear partial differential equations.

6.1 Governing equations

The heat and mass transfer is governed by the two second order non-linear partial differential equations (Luikov, 1966). Only one equation which describes heat conduction governed predominantly by temperature gradients was considered:

$$\varrho c_p \frac{\partial T}{\partial t} = k_x \frac{\partial^2 T}{\partial x^2} + k_y \frac{\partial^2 T}{\partial y^2}, \tag{33}$$

where k_x and k_y represent thermal conductivity (W/mK) in directions x and y of the cross-section of the beam, ϱ is density (kg/m^3), c_p specific heat (J/kgK) and T temperature (K). The second equation describes moisture diffusion governed by moisture potential and is not considered here.

The problem is complete when initial and boundary conditions are specified. The initial condition prescribes the temperature in the cross-section of the beam at the initial time $t = 0$

$$T(x, y, 0) = T_0(x, y). \tag{34}$$

The boundary conditions prescribe the heat flow on the exposed boundaries of a cross-section. Thus, the Neumann boundary conditions at the exposed surface are given by balancing heat

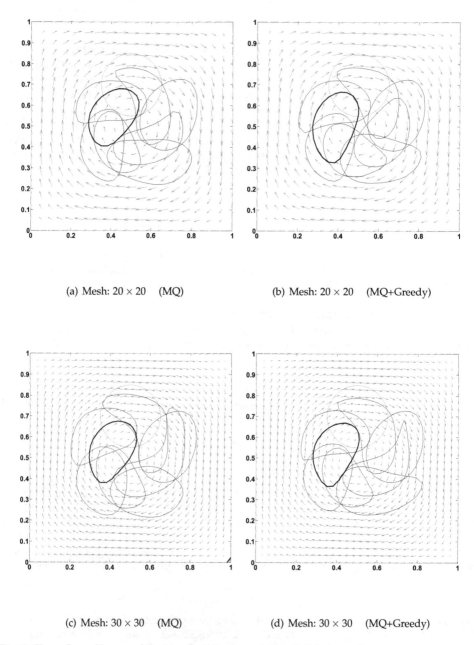

(a) Mesh: 20×20 (MQ)   (b) Mesh: 20×20 (MQ+Greedy)

(c) Mesh: 30×30 (MQ)   (d) Mesh: 30×30 (MQ+Greedy)

Fig. 5. Shear flow: Shapes of the circular interface at time $0, 0.5, 1, \ldots, 4$

conduction at the surface with the radiative and convective heat flux:

$$-k_x \frac{\partial T}{\partial x} e_{nx} - k_y \frac{\partial T}{\partial y} e_{ny} = h_c(T - T_A) + \varepsilon_R \sigma(T^4 - T_R^4), \tag{35}$$

where e_{nx} and e_{ny} are components of the normal to the boundary surface and h_c is convective heat transfer coefficient (W/m^{2}K). T_A is the ambient temperature. T_R is the temperature of the radiating surface, ε_R is the effective surface emissivity and σ is the Stefan-Boltzmann constant $\sigma = 5.671.10^{-8}$ W/m^{2}K^4.

6.2 Implicit discrete scheme

For the solution of eqn. (33) with the corresponding initial and boundary conditions, the RBF method is used. We consider the implicit time scheme of eqns. (33) and (35):

$$\varrho c_p \frac{T^{n+1} - T^n}{\triangle t} + k_x \frac{\partial^2 T^{n+1}}{\partial x^2} + k_y \frac{\partial^2 T^{n+1}}{\partial y^2} = 0, \tag{36}$$

$$-k_x \frac{\partial T^{n+1}}{\partial x} e_{nx} - k_y \frac{\partial T^{n+1}}{\partial y} e_{ny} - h_c(T^{n+1})$$
$$-\varepsilon_R \sigma((T^{n+1})^4 - T_R^4) = -h_c(T_A), \tag{37}$$

where $t_{n+1} = t_n + \triangle t$, T^{n+1} and T^n are the temperature at time t_{n+1} and t_n. The approximate solution is expressed as:

$$T(\mathbf{x}, t_{n+1}) = \sum_{j=1}^{N} \alpha_j^{n+1} \varphi_j(\mathbf{x}), \tag{38}$$

where α_j^{n+1}, $j = 1, ..., N$, are the unknown coefficients to be determined and $\varphi_j(\mathbf{x}) = \sqrt{(x - x_j)^2 + (y - y_j)^2 + c^2}$ are Hardy's multiquadrics functions (Hardy, 1971).
By substituting eqn. (38) into eqns. (36) and (37) and using factorization for the radiation term $(T^4 - T_R^4)$, we obtain:

$$\sum_{j=1}^{N} \left(\varrho c_p \frac{\varphi_j(\mathbf{x}_i)}{\triangle t} + k_x \frac{\partial^2 \varphi_j(\mathbf{x}_i)}{\partial x^2} + k_y \frac{\partial^2 \varphi_j(\mathbf{x}_i)}{\partial y^2} \right) \alpha_j^{n+1}$$
$$= \varrho c_p \frac{T^n(\mathbf{x}_i)}{\triangle t}, \quad i = 1, ..., N - N_B, \tag{39}$$

$$\sum_{j=1}^{N} \left(-k_x \frac{\partial \varphi_j(\mathbf{x}_i)}{\partial x} e_{nx} - k_y \frac{\partial \varphi_j(\mathbf{x}_i)}{\partial y} e_{ny} - h_c \varphi_j(\mathbf{x}_i) \right.$$
$$\left. -\varepsilon_R \sigma \left(\varphi_j(\mathbf{x}_i) - T_R \right) \left(\varphi_j(\mathbf{x}_i) + T_R \right) \left(\varphi_j^2(\mathbf{x}_i) + T_R^2 \right) \right) \alpha_j^{n+1}$$
$$= -h_c(T_A), \quad i = N - N_B + 1, ..., N, \tag{40}$$

where N_B and N present the number at boundary and all discretized points.

The system of nonlinear equations which result from the space discretization of a nonlinear PDEs were solved by Picard's methods.

6.3 Numerical examples

The results are tested on the one-dimensional case in the standard fire conditions, for which comparison is made with the results of one-dimensional charring rate models for wood presented in the literature (White & Nordheim, 1992). The same model is used to analyse a two-dimensional behaviour of timber beam exposed to fire from three sides.

One-dimensional charring

The charring rate of wood usually refers to the dimensional rate, e.g. millimetres per minute, at which wood transforms to char. Many factors are involved in wood charring. No completely satisfactorily charring model has yet been developed.

Since the material properties at elevated temperatures are difficult to obtain, constant material properties of the wood and char are used. The following data have been used:

$$T_0 = 20°C, \varrho = 370 \text{ kg/m}^3, k_{wood} = 0.12 \ k_{char} = 0.15 \text{ W/mK}, d = 0.3 \text{ m}$$
$$h_c = 22.5 \text{ W/m}^2, \epsilon_R = 0.9, c_{p,wood} = 1530 \text{ J/kgK}, c_{p,char} = 1050 \text{ J/kgK}.$$

Most known models suggest constant charring rates. In our case, we used a (White & Nordheim, 1992) non-linear empirical model for charring rate of eight different wood species. The comparison to the present model in the case of spruce is shown in Fig. 6.

In all empirical models it is assumed that the charring of woods starts instantaneously after exposure to fire. In reality, the charring is not immediate. In our model, the charring starts when the temperature of wood reaches the temperature of pyrolysis, which is around 300 °C. This temperature is reached nearly 3 minutes after the fire starts.

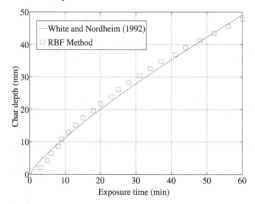

Fig. 6. Comparison of White and Norheim charring model to the present.

A two-dimensional charring

In the two-dimensional case, the formation of char in a timber beam exposed to the standard fire conditions (ISO 834, 1999) on three sides is considered. The upper surface is thermally isolated. The original beam cross-section is rectangular with dimensions 10×15 cm. The beam cross-section is discretized by the mesh of 10×10 points. Material properties are assumed to be the same as the one-dimensional case. The results of the simulation at 10 and 30 minutes after the exposure to fire are given in Fig. 7.

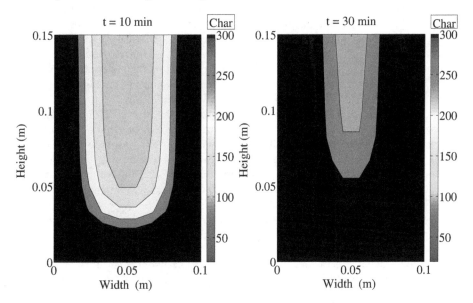

Fig. 7. Temperature distribution in the cross-section of spruce beam and the transformation of wood into char at 10 and 30 minutes calculated with the Kansa approach, relaxation parameter: 0.14.

7. Conclusions

Modelling of radionuclide migration, moving-boundary and wood charring were presented. The first example presents modelling of radionuclide migration through the geosphere using a combination of radial basis function methods in Eulerian coordinates with geostatistics.

In the case of radionuclide migration two steps of evaluations were performed. In the first step the velocities were determined from the pressure of the fluid p by solving the Laplace differential equation. In the second step the advection-dispersion equation was solved to find the concentration of the contaminant.

Comparison of the results between the average of contaminant concentrations show that the results are also very similar to the results obtained by finite difference methods (Fig. 3).

In the case of calculating the advection-dispersion equation we can conclude that the Kansa method could be an appropriate alternative to the FDM due to its simpler implementation.

The second example outlines our work done on an alternative approach to the conventional level set methods for solving two-dimensional moving-boundary problems. This approach is set up from MQ RBFs and the adaptive greedy algorithm. The examples suggest that the solution is more stable by employing the adaptive algorithm. Two examples are presented: oriented flow and shear flow (bubble starts circular but because of the velocity fields the bubble stretches). The results of the last example do not depend significantly on the usage of the adaptive greedy algorithm.

In the conventional level set methods, the level set equation is solved to evolve the interface using a capturing Eulerian approach. The solving procedure requires appropriate choice of the upwind schemes, reinitialization algorithms and extension velocity methods, which may require excessive amount of computational efforts. In this case we do not choose the reinitialization, because we try to control the smoothness of the moving boundary with appropriate choice of the type of MQ RBFs, shape parameters, time stepping schemes and greedy algorithm. The proposed alternative approach offers to use smaller computational grids, with no reinitialization in order to beat the upwind scheme. In the case of radionuclide migration modelling, the FDM FORTRAN program and the MQ FORTRAN program use comparable CPU times. Usually, fine discretization is applied, but this approch increases CPU time and can cause ill-conditioning. Even when solving two- or three-dimensional problems one tries to obtain the best accuracy with the least amount of CPU time. There are several methods used with RBF methods and other approaches that should be examined before attempting to solve two- or three-dimensional problems with many points, see (Scott & Kansa, 2009).

The last example shows the alternative approach for solving wood charring problems with the RBFs methods. Fig. 6 shows that presented approach and the model proposed by (White & Nordheim, 1992) produce comparable results. The same model was used to analyse a two-dimensional behaviour of timber beam exposed to fire from three sides. It shows that the results are comparable to the results obtained in literature (Schnabl & Turk, 2006).

8. References

Fredlund, B. (1993). Modelling of Heat and Mass Transfer in Wood Structures During Fire. *Fire Safety Journal*, Vol. 20, pp. 39-69.

Vrankar, L.; Turk, G. & Runovc, F. (2004). Modelling of radionuclide migration through the geosphere with radial basis function method and geostatistics. *Journal of the Chinese Institute of Engineers*, Vol.27, No. 4, pp. 455-462.

Vrankar, L., Turk, G. & Runovc, F. (2004). Combining the radial basis function Eulerian and Lagrangian schemes with geostatistics for modelling of radionuclide migration through the geosphere. *Computers and Mathematics with Applications*, Vol. 48, pp. 1517-1529.

Vrankar, L., Turk, G. & Runovc, F. (2005). A comparison of the effectiveness of using the meshless method and the finite difference method in geostatistical analysis of transport modelling. *Int J Comput Methods*, Vol. 2(2), pp. 149-166.

Wang, S. & Wang, M. Y. (2006). Radial basis functions and level set method for structural topology optimization. *International Journal for numerical methods in Engineering*, Vol. 65, pp. 2060-2090.

Ling, L. (2003). *Radial basis functions in scientific computing*, Ph. D. Thesis, Department of Mathematics, Simon Fraser University, Canada.

Kitanidis, P. K. & VoMvoris, E. G. (1983). A geostatistical approach to the inverse problem in groundwater modelling (steady state) and one-dimensional simulations. *Water Resources Research*, Vol. 19(3), pp. 677-690.

Deutsch, C. V. & Journel, A. G. (1998). *GSLIB Geostatistical Software Library and User's Guide*, Oxford University Press.

Olea, R. (1991). *Geostatistical Glossary and Multilingual Dictionary*. Oxford University Press, New York.

John, C. D. (1986). *Statistics and Data Analysis in Geology*. John Wiley & Sons, New York.

Atluri, S. N. & Shen, S. (2002). *The meshless local Petrov-Galerkin (MLPG) method*, Encino, CA: Tech Science Press. Vol. 191(21-22), pp. 2337-2370.

Li, G. & Aluru, N. R. (2002). Boundary cloud method: a combined scattered point/boundary integral approach for boundary-only analysis. *Comput Methods Appl Mech Eng*, Vol. 191(21-22), pp. 2337-2370.

Kansa, E. J. (1990). Multiquadrics - A scattered data approximation scheme with applications to computational fluid-dynamics. II: Solutions to parabolic, hyperbolic and elliptic partial differential equations. *Comput Math Appl 1990*, Vol. 19(8-9), pp. 147-161.

Hardy, R. L. (1971). Multiquadric equations of topography and other irregular surfaces. *J Geophys Res*, Vol. 176, pp. 1905-1915.

Wang, J. G. & Liu, G. R. (2002). On the optimal shape parameters of radial basis functions used for 2-D meshless methods. *Comput Meth Appl Mech Eng*, Vol. 191, pp. 2611-2630.

Wendland, H. (1995). Piecewise Polynomial, Positive Definite and Compactly Supported Radial Basis Functions of Minimal Degree. *Advances in Computational Mathematics*, Vol. 4, pp. 389-396.

Ling, L. & Kansa, E. J. (2004). Preconditioning for radial basis functions with domain decomposition methods. *Math Comput Modelling*, Vol. 40(13), pp. 1413-1427.

Kansa, E. J. (1990). Multiquadrics - A scattered data approximation scheme with applications to computational fluid-dynamics. I: Surface approximations and partial derivative estimates. *Comput Math Appl 1990*, Vol. 19(8-9), pp. 127-145.

Ling, L. & Hon, Y. C. (2005). Improved numerical solver for Kansa's method based on affine space decomposition. *Eng Anal Bound Elem*, Vol. 29(12), pp. 1077-1085.

Fornberg, B. & Wright, G. (2004). Stable computation of multiquadric interpolants for all values of the shape parameter. *Comput Math Appl*, Vol. 48(5-6), pp. 853-867.

Iske, A. (1994). *Characterization of conditionally positive definite functions for multivariable interpolation methods with radial basis functions. (Charakterisierung bedingt positiv definiter Funktionen für multivariate Interpolationsmethoden mit radialen Basisfunktionen.)*, Ph. D. Thesis, Göttingen: Math.-Naturwiss. FB, Univ. Göttingen.

Bear, J. (1972). *Dynamics of Fluids in Porous Media*, Elsevier, London.

Osher, S. & Fedkiw, R. (2003). *Level set methods and dynamic implicit surfaces.*, Applied Mathematical Sciences,Vol. 153, New York, NY: Springer.

Sethian, J. A. (1999). *Level set methods and fast marching methods. Evolving interfaces in computational geometry, fluid mechanics, computer vision, and materials science.*, Cambridge Monographs on Applied and Computational Mathematics, Vol. 3, Cambridge University Press.

Tsai, R. & Osher, S. (2003). Level set methods and their applications in image science. *Commun Math Sci*, Vol. 1(4), pp. 623-656.

Swiss Federal Institute of Technology (ETH) (2003). Short courses (Part IIB): Computational Multi-Fluid Dynamics (CMFD). Zurich, Switzerland, 24-28 March.

Vrankar, L., Kansa, E.J., Ling, L., Runovc, F. & Turk, G. (2010). Moving-boundary problems solved by adaptive radial basis functions. *Comput Fluids*, Vol. 39, pp. 1480-1490.

Luikov, A. V. (1966). *Heat and Mass Transfer in Capillary-porous Bodies*, Pergamon Press, Oxford.

Hardy, R. L. (1971). Multiquadric equations of topography and other irregular surfaces. *J. Geophys Res.*, Vol. 176, pp. 1905-1915.

White, R. H. & Nordheim, E. V. (1992). Charring rate of wood for ASTM E 119 exposure. *Fire Technology*, Vol. 28(1), pp. 5-30.

ISO 834 (1999) *Fire-resistance test-Elements of building construction-Part 1. General requirements. ISO 834-1*, International organization for standardization, Geneva, Switzerland.

Schnabl, S. & Turk, G. (2006). Coupled heat and moisture transfer beams exposed to fire. *WCTE 2006 - 9th World Conference on Timber Engineering - Portland, OR, USA.*

Scott, A. S. & Kansa, E. J. (2009). *Multiquadric Radial Basis Function Approximation Methods for the Numerical Solution of Partial Diferential Equations*, Advances in Computational Mechanics, Vol. 2, 2009, ISSN: 1940-5820.

8

Simulation of Acoustic Sound Produced by Interaction Between Vortices and Arbitrarily Shaped Body in Multi-Dimensional Flows by the Vortex Method

Yoshifumi Ogami
Ritsumeikan University
Japan

1. Introduction

One of the most important environmental and/or engineering issues at present is the reducing aerodynamic noise that is produced from, for example, the body surfaces of cars; parts of cars, such as sideview mirrors, windshield wipers, and roof racks; pantographs of bullet trains; narrow spaces between cars of bullet trains; and fans in air-conditioners.

In order to analyze aerodynamic sound for practical engineering use, the decoupled solution method — which treats the flow field and acoustic field separately — has been employed. In this method, the flow field involves an inviscid incompressible fluid that yields a sound source, and the acoustic field is analyzed by combining this source with wave equations, such as the Lighthill equation. Powell (1964) and Howe (2003) showed that this sound source is represented by a vorticity vector. Therefore, aerodynamic sound is considered to originate from the unsteady motion of vorticity; that is, generation and deformation of vorticity, accelerated motion of vorticity, interaction of vorticity, and interference of obstacles. This chapter mainly focuses on the simulation of vortex sounds created by obstacle interference in multi-dimensional flows through the simple but powerful vortex method.

Using the compact Green's function, which is based on the compact assumption that the wavelength of sound is large compared to the dimensions of the solid body, Howe (2003) studied the vortex sound for one of the most fundamental cases of interference: namely, where one vortex line passes near an obstacle. In this study, the vortex movement was calculated based on the potential flow of the inviscid and incompressible fluid so that the computational load and time would be very small. The fundamental aspects of the sound produced by an interaction between vorticity and obstacle are well understood. However, the obstacles are limited to two-dimensional simple shapes because the conformal mappings to create the obstacles have to be known.

Therefore, this chapter presents a generalization of Howe's method to treat two-dimensional shapes with unknown conformal mappings and three-dimensional obstacles in as simple manner as possible. This generalization will help in analyzing the sound produced by vorticity and an arbitrarily shaped body with less computational effort for studying the effect of the body shape on the aerodynamic sound, controlling the aerodynamic sound, etc. To calculate the flow fields in two and three dimensions, we employ the vortex method (panel method), which can directly and easily deal with an arbitrarily shaped body with vortices in non-linear motion.

The vortex method has been used to analyze aerodynamic sounds through the various equations or methods listed below:

1. Sharland's equation that relates the lift coefficient to sound power (Sharland, 1964).
2. Curl's equation that calculates the aerodynamic sound from the pressure fluctuations on the surface of a body (Curle, 1955).
3. The Ffowcs Williams–Hawkings equation that treats the aerodynamic sound of rotating wings (Ffowcs Williams & Hawkings, 1969, Huberson et al., 2008).
4. The asymptotic matching method and panel method (Kao, 2002).
5. Howe's method with the compact Green's function (Howe, 2003).

Of these methods, Howe's method provides a simple and direct relation between the vortex sound and vortex. Ogami and Akishita (2010) applied Howe's method to the aerodynamic sound produced by the interaction of a flat plate and vortices in non-linear motion. The relation between the acoustic pressure and vortices is directly expressed in a simple manner so that the effects of the vortices in non-linear motion on the acoustic sound can be studied easily with less computational effort than the other methods. As stated before, one of the purposes of this chapter is to generalize Howe's method to treat two-dimensional shapes with unknown conformal mappings and three-dimensional obstacles; the details are explained in Section 2. The idea for calculating the acoustic pressure of arbitrarily shaped obstacles without conformal mappings is to use three sets of bound vortices that represent the obstacles in uniform flows of speed 1 in the x and y directions in two dimensions as well as in the z-direction in three dimensions. These bound vortices are obtained by the panel method in advance.

To examine the validity and accuracy of our method, we compared our solutions with those of Howe for the acoustic pressure produced by the interaction of an incident vortex and circular cylinder, and we present our findings in Section 3.

Howe (1976) showed that *the imposition of a Kutta condition at the trailing edge leads to a complete cancellation of the sound generated when frozen turbulence (the turbulence is modeled by a line vortex convected in a mean flow) convects past a semi-finite plate and to the cancellation of the diffraction field produced by the trailing edge in the case of an airfoil of compact chord.* However, his results were based on the linear assumption that the strength of the incident vortex is small enough so that its path is almost linear and that the wake vortices are also swept downstream at the mean stream velocity. Since our method is applicable to non-linear simulations where the wake vortices are swept at velocities induced by other vortices, both an airfoil (Section 3) and semi-finite plate (Section 4) were considered with fully non-linear calculations. The shedding of wake vortices was found to not cancel the sound but has the extra effect of producing sound due to the accelerated movements of the wake vortices.

Our findings were also compared with Kao's solutions by the asymptotic matching method, as presented in Section 4. Our solutions are considerably simpler and easier to code than those of Kao's analysis.

Section 5 discusses the sound produced by the vortex motion near a half-plane to study the effect of shedding vortices from the edge on the sound. Section 6 presents the study of a three-dimensional rectangular solid for examining the effect of three-dimensionality for an incident vortex on the vortex sound. The shedding of wake vortices was not considered.

In this study, two types of vortices were used: the free vortex and the bound vortex. The movements of the free vortices, which were the sources of the sound, were calculated on the basis of the potential flow of the inviscid and the incompressible fluid as stated before. The bound vortices were used for representing an arbitrarily shaped body. This body can be created in the potential flow by adjusting the strengths of the bound vortices, which were arranged on the virtual panels placed along the body, under the non-permeable condition that the stream line does not penetrate this body.

2. Acoustic pressure for arbitrarily shaped bodies with free vortices in non-linear motion

As stated in the Introduction, Howe's solutions (2003) are limited to body shapes whose conformal mappings are known. In this section, we derive the relations between the acoustic pressures in the two-, quasi three-, and three-dimensions and the vortices for interactions with arbitrarily shaped obstacles.

2.1 Two dimensions
Consider the following wave equation:

$$\left(\frac{1}{c^2}\frac{\partial^2}{\partial t^2} - \nabla^2\right)\frac{p(x,t)}{\rho} = \mathrm{div}(\omega \wedge v) \tag{1}$$

where c is the sound velocity in the fluid, $p(x,t)$ is the acoustic pressure at the observation location x and time t, ρ is the fluid density, and ω is the vortex vector swept at velocity vector v. Using this equation, Howe derived the acoustic pressure $p(x,t)$ produced by a vortex of strength Γ located at (x_{01}, x_{02}) in a two-dimensional flow as follows:

$$p(x,t) = \frac{-\rho\Gamma x_j}{2\pi\sqrt{2c}\,|x|^{\frac{3}{2}}}\frac{\partial}{\partial t}\int_{-\infty}^{t-|x|/c}\left\{\frac{dx_{01}}{d\tau}\frac{\partial Y_j}{\partial y_2} - \frac{dx_{02}}{d\tau}\frac{\partial Y_j}{\partial y_1}\right\}\frac{d\tau}{\sqrt{t-\tau-|x|/c}} \tag{2}$$

where Y_j is the component of the Kirchhoff vector Y, which depends on the shape of the body and is considered to be a velocity potential with unit speed in the j direction. Note that the coordinates x and x_j denote the position of the observer, whereas (x_{01}, x_{02}) indicate the position of the incident vortex.

For the vortex method (panel method), Eq. (2) can be rewritten as explained below.

In a two-dimensional potential flow with unit speed in the x direction and an arbitrarily shaped body represented by bound vortices of strength Γ_{Bi}^X (the subscript X denotes the the x direction, and B_i indicates the i-th bound vortex), the complex potential, F, and velocity potential, ϕ, are given by

$$F = Uz + \sum_{i=1}^{N}\frac{i\Gamma_{Bi}^X}{2\pi}\log(z - z_i) \tag{3}$$

$$\phi = Y_1 = Ux - \sum_{i=1}^{N}\frac{\Gamma_{Bi}^X}{2\pi}\theta_i \qquad (\theta_i \text{ is argument of } z - z_i) \tag{4}$$

Similarly, for a potential flow with unit speed in the y direction and bound vortices of strength Γ^Y_{Bi}, the equations are

$$F = Uze^{-i\frac{\pi}{2}} + \sum_{i=1}^{N} \frac{i\Gamma^Y_{Bi}}{2\pi} \log(z - z_i) \tag{5}$$

$$\phi = Y_2 = Uy - \sum_{i=1}^{N} \frac{\Gamma^Y_{Bi}}{2\pi} \theta_i \tag{6}$$

By using Eqs. (2), (4) and (6), we get the equation below, which directly relates the acoustic pressure to the vortices.

$$p(x,t) = \sum_{j=1}^{M} \frac{\rho_0 \Gamma_{Fj} x_1}{2\pi\sqrt{2c_0} |x|^{\frac{3}{2}}} \frac{\partial}{\partial t} \int_{-\infty}^{t-|x|/c_0} \left\{ u_{Fj} \sum_{i=1}^{N} \frac{\Gamma^X_{Bi}}{2\pi} \frac{x_j - x_i}{\left(x_j - x_i\right)^2 + \left(y_j - y_i\right)^2} + \right.$$

$$\left. v_{Fj} \left(1 + \sum_{i=1}^{N} \frac{\Gamma^X_{Bi}}{2\pi} \frac{y_j - y_i}{\left(x_j - x_i\right)^2 + \left(y_j - y_i\right)^2} \right) \right\} \frac{d\tau}{\sqrt{t - \tau - |x|/c_0}}$$

$$+ \sum_{j=1}^{M} \frac{\rho_0 \Gamma_{Fj} x_2}{2\pi\sqrt{2c_0} |x|^{\frac{3}{2}}} \frac{\partial}{\partial t} \int_{-\infty}^{t-|x|/c_0} \left\{ u_{Fj} \left(-1 + \sum_{i=1}^{N} \frac{\Gamma^Y_{Bi}}{2\pi} \frac{x_j - x_i}{\left(x_j - x_i\right)^2 + \left(y_j - y_i\right)^2} \right) + \right.$$

$$\left. v_{Fj} \sum_{i=1}^{N} \frac{\Gamma^Y_{Bi}}{2\pi} \frac{y_j - y_i}{\left(x_j - x_i\right)^2 + \left(y_j - y_i\right)^2} \right\} \frac{d\tau}{\sqrt{t - \tau - |x|/c_0}} \tag{7}$$

where Γ_{Fj} and (u_{Fj}, v_{Fj}) represent the j-th free vortex and its velocity components. The bound vortices Γ^X_{Bi} and Γ^Y_{Bi} are obtained in advance by the panel method. The first term of the right-hand side of Eq. (7) is called the acoustic pressure of the *drag dipole* (*suction dipole*), and the second term is that of the *lift dipole*. Note that an integral from $-\infty$ to $t - |x|/c_0$ and a time-derivative remain in the solution (7). These operations are numerically calculated in this chapter.

2.2 Quasi three dimensions
Howe derived the acoustic pressure $p(x,t)$ produced by vortex vector ω located at y in a three-dimensional flow as follows.

$$p(x,t) = \frac{-\rho_0 x_j}{4\pi c_0 |x|^2} \frac{\partial}{\partial t} \int (\omega \wedge v) \left(y, t - \frac{|x|}{c_0} \right) \cdot \nabla Y_j(y) d^3y \tag{8}$$

Following Howe's analysis, we assume that a line vortex that interacts with a body remains rectilinear as when it is introduced in the flow. This enables us to use the two-dimensional velocity potential for this analysis, even though the body and the flow are three-dimensional. Therefore, the solutions for this section are considered to be quasi three-dimensional. Fully three-dimensional analysis, where the vortex changes its shape, is introduced in the next section.

Simulation of Acoustic Sound Produced by Interaction Between Vortices and Arbitrarily Shaped Body in Multi-Dimensional Flows by the Vortex Method

161

When the span of the body in the z direction is L (Fig. 1), the components of the Kirchhoff vector are

$$Y_1 = y_1 - \sum_{i=1}^{N} \frac{\Gamma_{Bi}^X}{2\pi} \theta_i$$

$$Y_2 = \begin{cases} y_2 - \sum_{i=1}^{N} \frac{\Gamma_{Bi}^Y}{2\pi} \theta_i & |y_3| < \frac{1}{2}L \\ y_2 & |y_3| > \frac{1}{2}L \end{cases}$$

$$Y_3 = y_3$$

(9)

Using the Dirac delta function δ, the vortex vector is expressed as

$$\omega = (0,0,\Gamma_{Fj}\delta(y_1 - \int u dt)\delta(y_2 - \int v dt))$$

(10)

where Γ_{Fj} is the strength of the j-th line vortex.

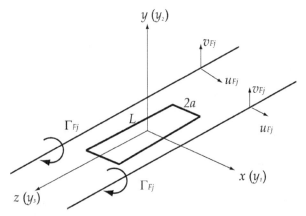

Fig. 1. Coordinate system.

Eqs. (9) and (10) are substituted into Eq. (8) to obtain

$$
\begin{aligned}
p(x,t) = \sum_{j=1}^{M} & \frac{\rho_0 x_1 L}{4\pi c_0 |x|^2} \frac{\partial}{\partial t} \Gamma_{Fj} \left\{ v_{Fj} \left(1 + \sum_{i=1}^{N} \frac{\Gamma_{Bi}^X}{2\pi} \frac{y_j - y_i}{(x_j - x_i)^2 + (y_j - y_i)^2} \right) \right. \\
& \left. + u_{Fj} \left(\sum_{i=1}^{N} \frac{\Gamma_{Bi}^X}{2\pi} \frac{x_j - x_i}{(x_j - x_i)^2 + (y_j - y_i)^2} \right) \right\} \\
+ & \frac{\rho_0 x_2 L}{4\pi c_0 |x|^2} \frac{\partial}{\partial t} \Gamma_{Fj} \left\{ v_{Fj} \left(\sum_{i=1}^{N} \frac{\Gamma_{Bi}^Y}{2\pi} \frac{y_j - y_i}{(x_j - x_i)^2 + (y_j - y_i)^2} \right) \right. \\
& \left. - u_{Fj} \left(1 - \sum_{i=1}^{N} \frac{\Gamma_{Bi}^Y}{2\pi} \frac{x_j - x_i}{(x_j - x_i)^2 + (y_j - y_i)^2} \right) \right\}
\end{aligned}
$$

(11)

where the first term of the right-hand side of Eq.(11) is called the acoustic pressure of *drag dipole* (*suction dipole*) and the second term that of *lift dipole*. As in the two-dimensional analysis, the acoustic pressure is expressed with the free vortices, and with the bound vortices Γ_{Bi}^X and Γ_{Bi}^Y that are to be obtained in advance by the panel method. Note that the integral that appeard in Eq. (7) does not appear in Eq.(11) but the time-derivative remains.

2.3 Three dimensions

In three dimensions, the acoustic pressure can be calculated directly if the bound vortices Γ_{Bi}^j, representing the body within the unit flow in the j-direction (j = x, y, and z), are obtained in advance. In terms of the velocity vector (u_{Bj}, v_{Bj}, w_{Bj}) produced by the bound vortices Γ_{Bi}^j and unit flow in the j-direction, the components of the Kirchhoff vector are written as

$$\nabla Y_j(y) = (u_{Bj}, v_{Bj}, w_{Bj}) \tag{12}$$

Therefore, the following term in Eq. (8) can be expressed as

$$(\boldsymbol{\omega} \wedge v) \cdot \nabla Y_j(y) = (\omega_2 w - \omega_3 v, \ \omega_3 u - \omega_1 w, \ \omega_1 v - \omega_2 u) \cdot (u_{Bj}, v_{Bj}, w_{Bj}) \tag{13}$$

Using the free vortices (incident vortex and wake vortices) numbered with i, Eq. (8) is expressed as

$$p(x,t) = \frac{-\rho_0 x_j}{4\pi c_0 |x|^2} \frac{\partial}{\partial t} \sum_{i=1}^{N} (\boldsymbol{\Gamma}_i \wedge v) \cdot \nabla Y_j \tag{14}$$

where $\boldsymbol{\Gamma}_i$ is the strength vector of the vorticity vector $\boldsymbol{\omega}_i$ in volume $d^3 y$.

3. Panel method and circular cylinder

3.1 Panel method for acoustic pressure

As shown in Fig. 2, the obstacle in the two-dimensional flow is represented by the discrete bound vortices of strength Γ_{Bi} (dots), and the strength is determined by the non-permeable condition that the components of the fluid velocities normal to the body surface (to the points between the bound vortices (stars) in this study) are zero. The total strength of the bound vortices is assumed to be zero. When there is more than one free vortex in the flow, the strength of the bound vortices varies with the influence of the moving free vortices, so the strength of the bound vortices must be calculated for every time step. On the other hand, the strengths of the bound vortices Γ_{Bi}^X and Γ_{Bi}^Y in Eqs. (7) and (11) are only obtained once because there is no need to consider the influence of the free vortices for these bound vortices.

3.2 Comparison with Howe's solutions

The acoustic pressure produced by the interaction of a free vortex (incident vortex) and circular cylinder with radius 1 in a uniform flow of velocity 1 was considered. The cylinder was represented by 80 bound vortices. An incident vortex of strength -4π was placed at time 0 at the position x = -10, y = -0.7. The Adams-Bashforth method of second order accuracy was used to update the position of the incident vortex at time steps of 0.01. The acoustic pressure was calculated by using Eq. (7) for the two-dimensional analysis. The integral from $-\infty$ to $t-|x|/c_0$ in Eq. (7) was evaluated numerically from a negative value of sufficiently large absolute value to the present time.

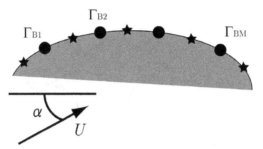

Fig. 2. Positions of bound vortices (dots) and for non-permeable condition (stars).

Figure 3 compares the paths of the incident vortex according to Howe's solution (p. 196, Howe, 2003) and the present method. The incident vortex first placed in the third quadrant at $x = -10$, $y = -0.7$ exited this quadrant and entered the second quadrant as it approached the cylinder, flowed along the cylinder surface, and was swept downstream. Although this movement was caused by the effect of the mirror image vortex, the bound vortices placed along the cylinder surface had this effect without the placement of an image vortex inside the cylinder. The paths of both solutions shown in Fig. 3 agreed well, with an average relative error of 0.05%.

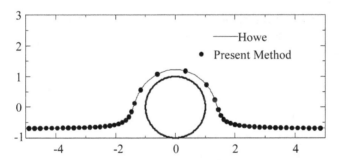

Fig. 3. Path of incident vortex.

The average relative error is defined by Eq.(15),

$$\text{Error} = \frac{1}{n}\sum_{i=1}^{n}\left|\frac{S_{Howe}^i - S_P^i}{S_{MAX}}\right|\times 100\% \qquad (15)$$

where S_{Howe}^i and S_P^i are Howe's solution and that of the present method, respectively, at the i-th time step, $|S_{MAX}|$ is the maximum absolute value during the period of calculation, and n is the number of solutions obtained during this period.

Figure 4 compares the acoustic pressures of the drag dipole and lift dipole, as calculated by Howe's method (solid and broken lines) and the present method (dots). The average relative error was 0.3%. According to Howe (2003), the time was set to zero when the incident vortex passed the position $x = 0$. The parameters in Fig. 4 are $V = \Gamma / (2\pi a)$ and $a = 1$. The acoustic pressures of the drag and lift dipoles are non-dimensionalized by $k\sin\Theta$ and $k\cos\Theta$, respectively, where $k = \rho_0 V^2 \sqrt{M} \sqrt{a/|x|}$, $M = V/c_0$, and $\Theta = \tan^{-1}(x_1/x_2)$.

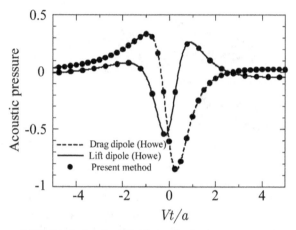

Fig. 4. Drag dipole and lift dipole for the circular cylinder.

Next, to determine the convergence and accuracy of the present method, the acoustic pressure by the drag dipole was calculated while changing the bound vortex number N and time step Δt. Figure 5 shows that for $N > 40$, the average relative error decreased with decreasing time step, and the difference between $N = 80$ and 120 was not noticeable; thus the solutions were considered to converge at $N \geq 80$. The error of the lift dipole had a similar tendency, so that figure is omitted.

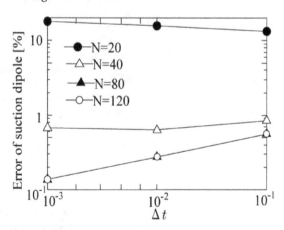

Fig. 5. Averaged relative error for drag dipole.

Figure 6 shows the average relative error for the paths of the incident vortices. The solutions converged with a smaller numbers of bound vortices ($N \geq 40$) than the acoustic pressure in Fig. 5 ($N \geq 80$).

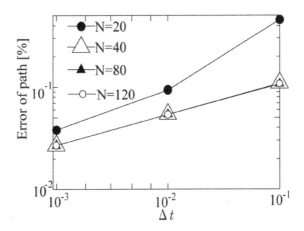

Fig. 6. Averaged relative error for path of the incident vortex.

For the simulations of engineering applications, multiple vortices passing the body and produced on the body surface have to be considered. The acoustic pressure is easily obtained from Eqs. (7) and (11) by summing the acoustic pressures produced by each vortex passing the body. Therefore, although the solutions shown in Fig. 4 are very elemental, they are important for studying the fundamental characteristics of vortex sound.

4. NACA0012 airfoil

4.1 Without wake vortices
As stated before, Howe reported that the wake vortex shed from the trailing edge of an airfoil cancels the acoustic pressure when the Kutta condition is employed. To verify this cancellation, we studied the effect of the wake vortices shed from the NACA0012 airfoil on the acoustic pressure. The airfoil was placed in a uniform flow of velocity 1 and was represented by 160 bound vortices; its leading edge lay at $x = 0$, and the trailing edge was at $x = 1$. The angle of attack was set to zero. The incident vortex of strength $\Gamma = -0.4\pi$ started at the position $x = -10$, $y = 0$. The time step was set to 0.01.

With the Kutta condition that the flow separates from the trailing edge, wake vortices were produced by the influence of the incident vortex even when the angle of attack was zero. We neglect the production of the wake vortices in this section and consider it in the next section. In addition to the Kutta condition, the total strength of the bound vortices was set to zero for the former case, and the total strength of the bound and wake vortices was set to zero for the latter case.

Figure 7, which is enlarged in the y direction, shows the path of the incident vortex when the production of the wake vortices was not considered. The incident vortex started at $x = -10$, $y = 0$, and as it approached the airfoil it was swept upward by the effect of the mirror image vortex. The path broke near the trailing edge, around positions D and E in Fig. 7, because the velocity at the trailing edge was very large (without the Kutta condition, the velocity in the analytical solution becomes infinity). The letters A–E indicate the positions of the incident vortex when its acceleration, shown in Fig. 9, reached the peak value, and the numbers in parentheses in Fig. 7 are the times of the peak.

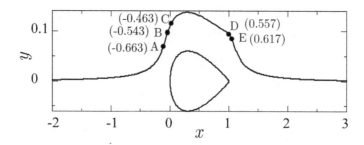

Fig. 7. Path of incident vortex around the NACA0012 airfoil.

Figure 8 shows the acoustic pressure calculated (a) by Eq. (7) for two-dimensional analysis and (b) calculated by Eq. (11) for quasi three dimensions. The time was set to zero when the incident vortex passed the center of the airfoil ($x = 0.5$) and was non-dimensionalized as Ut/a ($U = 1$ and $a = 0.5$). The acoustic pressures of the drag dipole and lift dipole for two dimensions were non-dimensionalized by $k\sqrt{M}\sin\Theta\sqrt{a/|x|}$ and $k\sqrt{M}\cos\Theta\sqrt{a/|x|}$, respectively, and those for quasi three dimensions were non-dimensionalized by $kM\sin\Theta L/|x|$ and $M\cos\Theta L/|x|$, respectively, where $k = \rho_0\Gamma U/(4\pi a)$ and $M = U/c_0$.

As shown in Fig. 8(a), when $Ut/a < -0.52$, the acoustic pressure of the drag dipole (broken line) increased and that of the lift dipole decreased (solid line) as the incident vortex approached the leading edge of the airfoil. At around $Ut/a = -0.52$, when the incident vortex passed the leading edge ($x = 0$), the former pressure reached the local maximum value, and the latter reached the local minimum value. The lift dipole then continued to increase due to the increase in acceleration du/dt of the incident vortex (see Fig. 9 for du/dt and dv/dt). Near $Ut/a=0.57$, when the incident vortex passed the trailing edge, both pressures had local minima because the acceleration of the incident vortex also had a local minimum. These minima occurred because the path of the vortex broke near the trailing edge, as shown in Fig. 7.

As shown in Figs. 8(a) and (b), the dipoles behaved similarly in quasi three dimensions and in two dimensions. The major difference was that when the incident vortex passed the trailing edge at around $t = 0.5$, the acoustic pressures for both dipoles reached local minima in quasi three dimensions and two dimensions, while afterwards they reached local maxima in quasi three dimensions only. A peak at the time corresponding to the local maximum was observed in the acceleration dv/dt of the incident vortex, as shown in Fig. 9. In general, the acoustic pressure resembled the acceleration more for the quasi three dimensions than the two dimensions. The stronger resemblance occurs because the solution was obtained by integration and derivation about the time for Eq. (7) in two dimensions while only derivation about time for Eq. (11) was used in quasi three dimensions.

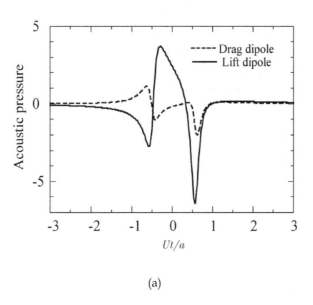

(a)

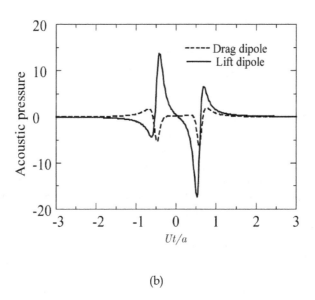

(b)

(a) Two dimensions; (b) Quasi three dimensions

Fig. 8. Drag and lift dipoles for NACA0012.

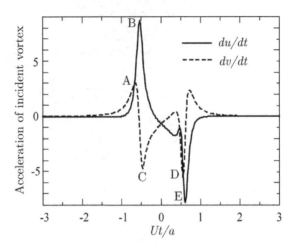

Fig. 9. Acceleration of incident vortex for two- and quasi three-dimensional analyses.

4.2 With wake vortices in non-linear motion

As stated before, Howe reported that the wake vortex shed from the trailing edge of an airfoil cancels the acoustic pressure when the Kutta condition is employed. However, his analysis was based on a linear assumption. To determine whether linearity is valid for more realistic flows with non-linear vortex movement, we conducted simulations considering the production of wake vortices that move with the velocities induced by each other. The wake vortices were produced one by one at every time step near the edge (i.e., $x = 1 + U\Delta t$ and $y = 0$), and the strength is determined by the Kutta condition, which set the strength of the bound vortex at the trailing edge at zero. The acoustic pressure was calculated by using Eq. (11) for quasi three dimensions.

Figure 10 compares the acoustic pressures with and without the wake vortices (solid and broken lines, respectively). Before the incident vortex reached the trailing edge ($Ut/a < 0.63$), the peak values for the acoustic pressures of the drag and lift dipoles were smaller when the wake vortices were considered than when they were not considered. The smaller size occurs because the wake vortices decrease the acceleration of the incident vortex (Fig. 11).

After the incident vortex reached the trailing edge, the acoustic pressure without the wake vortices converged to zero, while that with the wake vortices did not, due to the continuing unsteady movements of the wake vortices.

Figure 12(a) shows the wake vortices at time 7.5 when the incident vortex entraining the wake vortices passed the coordinates $x = 4.4$, $y = -0.17$. Figure 12(b) shows the path of the incident vortex. The movement looks like "hopping," and this continued to produce the acoustic pressure. When the time is very large and the position of the incident vortex is far enough from the object, the term for the incident vortex in Eq. (11) becomes

$$p(x,t) = \frac{\rho_0 L\Gamma x_1}{4\pi c_0 |x|^2} \frac{\partial v}{\partial t} - \frac{\rho_0 L\Gamma x_2}{4\pi c_0 |x|^2} \frac{\partial u}{\partial t} \qquad (16)$$

Simulation of Acoustic Sound Produced by Interaction Between Vortices and Arbitrarily Shaped Body in Multi-
Dimensional Flows by the Vortex Method
169

Because the accelerations $\partial u/\partial t$ and $\partial v/\partial t$ remain sound is produced even when the incident vortex flows far away from the airfoil in the potential theory with the compact assumption. However, in a physical flow, the vortex is diffused by the viscosity or dissipated by the turbulence, so the sound disappears.

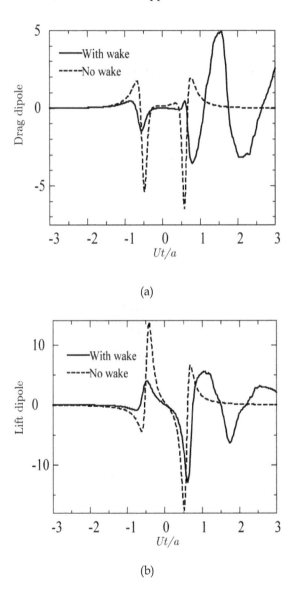

(a)

(b)

(a) Drag dipole; (b) Lift dipole

Fig. 10. Drag and lift dipoles with and without wake vortices.

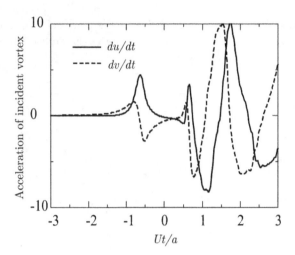

Fig. 11. Acceleration of the incident vortex with wake vortices.

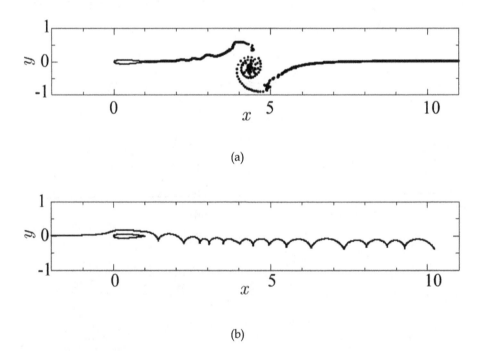

(a)

(b)

(a) Wake votices at t = 7.5; (b) Path of incident vortex

Fig. 12. (a) Wake vortices and (b) path of incident vortex.

Simulation of Acoustic Sound Produced by Interaction Between Vortices and Arbitrarily Shaped Body in Multi-
Dimensional Flows by the Vortex Method
171

Figure 13 shows the strength of the wake vortex produced at each time step. Over most of the simulation, the strength was two orders of magnitude smaller than that of the incident vortex $\Gamma = -0.4\pi \approx -1.26$, and its maximum was 0.052, which is about 4% that of the incident vortex.

Furthermore, to examine the production mechanism of the acoustic pressure, the simulations were conducted with the incident vortex strength reduced to one-tenth of its previous value: that is, $\Gamma = -0.04\pi \approx -0.126$.

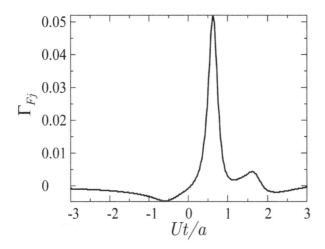

Fig. 13. Strength of wake vortices shed from trailing edge.

Figure 14 shows the acoustic pressures, with and without the wake vortices, for the drag and lift dipoles. Even though the strength of the incident vortex was decreased to one-tenth, the pressure of the drag dipole did not greatly differ from that shown in Fig. 10, while the lift dipole took a large peak value of -56.5 at time 0.75. This peak was not observed in the acceleration of the incident vortex shown in Fig. 15, so this peak was not affected by the acceleration of the incident vortex. On the other hand, the time of this peak coincides with that of the strength of wake vortices shed from the edge shown in Fig. 16. The peak value was almost 0.02, which was 16% of the incident vortex and relatively larger than the 4% in the previous simulation. The incident vortex passed the trailing edge at around time = 0.77, and the largest wake vortex was produced. This shows that the production of the wake vortex strongly affected the acoustic pressure and the acceleration of the incident vortex.

The results of our simulations show that the wake vortex shed from the trailing edge of an airfoil does not cancel the acoustic pressure. Before the incident vortex reaches the trailing edge, the pressure is decreased because of the effect of the wake vortices. However, afterwards, the movements of the wake vortices keep producing the large pressure, and the production of the wake vortices also increases the acoustic pressure.

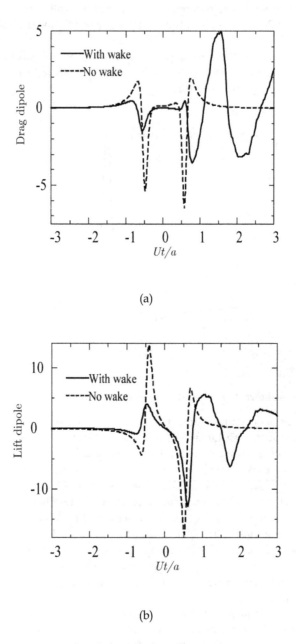

(a)

(b)

(a) Drag dipole; (b) Lift dipole.

Fig. 14. Drag and lift dipoles for $\Gamma = -0.04\pi$

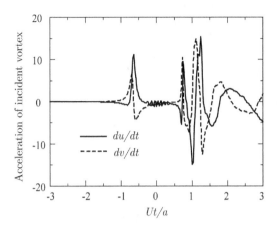

Fig. 15. Acceleration of incident vortex with the effect of wake vortices.

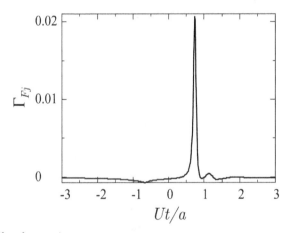

Fig. 16. Strength of wake vortices.

The reason for the disagreement may be because in Howe's analysis (1976), the wake vortices were arranged on the x axis from the trailing edge to infinity, and their strength was given by a periodic function. On the other hand, in our simulations, the wake vortices were not on the x axis but moved freely with nonlinear interaction of the vortices, and their strength was not periodic.

4.3 Comparison with asymptotic matching method

Kao (2002) studied the acoustic pressure by using the *asymptotic matching method*, which treats the near field as an incompressible fluid and the far field as a compressible fluid governed by the classical acoustic equation. The solutions in the two regions are matched asymptotically. Compared to the workload of Kao's analysis to obtain solutions, our solutions (7) and (11) are considerably simpler and easier to code.

Figure 17 compares the acoustic pressures produced by the interaction of an incident vortex, the wake vortices, and the NACA0012 airfoil as calculated by Kao (dots), the present method for quasi three dimensions (solid line), and that for two dimensions (broken line). The incident vortex of strength ± 0.1 was placed at $x = -5.0$, $y = 0.1$, and it passed above the NACA0012 airfoil located at $x = 0\text{-}1$. The production of the wake vortices was taken into account with the Kutta condition. Note that the strength of the incident vortex was so small that rolling up of the wave vortices was not observed, although the figure is not shown here.

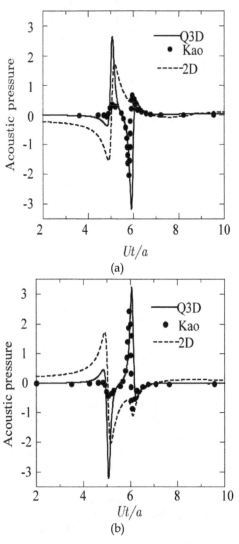

(a) $\Gamma = -0.1$; (b) $\Gamma = 0.1$

Fig. 17. Comparison with the asymptotic matching method.

Although Kao's analysis was two-dimensional and we used the compact assumption that the wavelength of sound is large compared to the dimension of the solid body, the solutions by Kao agreed very well with those for the present method in quasi three dimensions. In both Kao's and our simulations, the peak was observed when the incident vortex passed the leading edge at around $Ut/a = 5$ and the trailing edge at around $Ut/a = 6$.

5. Sound produced by vortex motion near a half-plane

Howe (2003) studied the sound produced by the vortex motion near a half-plane and concluded that *sound produced by the shed vortex tends to cancel the edge-generated sound attributable to the incident vortex* Γ *alone*. However, his analysis was qualitative explanation because he did not calculate the sound considering the wake vortices with nonlinear interactive movements. Therefore, we conducted simulations considering these movements. First, we considered the flow in the ζ-plane given by the complex function for the incident vortex of strength Γ located at $\zeta = \zeta_0$, its image vortex at $\zeta = \zeta_0^*$, and the wake vortices of strength Γ_m given by

$$F = -\frac{i\Gamma}{2\pi}\log(\zeta - \zeta_0) + \frac{i\Gamma}{2\pi}\log(\zeta - \zeta_0^*) + \sum_{m=1}^{N}\frac{i\Gamma_m}{2\pi}\left(\log(\zeta - \zeta_m) - \log(\zeta - \zeta_m^*)\right) \qquad (17)$$

where ζ^* is the complex conjugate of ζ. Note that the total strength of the vortices is zero. The flow about a half-plane was then obtained by transforming the flow given by Eq. (17) into the z-plane by the following conformal mapping:

$$\zeta = i\sqrt{z} \qquad (18)$$

The wake vortices were shed from the edge at each time step, and the strength was determined by the Kutta condition that removes an infinite velocity at the edge. Considering this, strength Γ_N of the N-th wake vortex was calculated by the recurrence formula

$$\Gamma_N = -\frac{\sum_{j=1}^{N-1}\Gamma_j\left(\dfrac{(\psi_j)}{(\xi - \xi_j)^2 + (\psi_j)^2}\right)}{\dfrac{(\psi_N)}{(\xi - \xi_N)^2 + (\psi_N)^2}} \qquad (19)$$

where ξ_j and ψ_j are the coordinates of the j-th free vortex (including the incident vortex) in the ζ-plane. The acoustic pressure can be calculated by applying Eq. (6.2.9) in Howe (2003) to multiple vortices.

The incident vortex of strength $\Gamma = 1$ was located at $x = -20$, $y = 0.4$, and the time step was set as 0.05. We assumed that the wake vortices were created near the edge of the half-plane (i.e., $x = 0.01$, $y = 0$) at every time step. The exact position of the edge was $x = 0$ and $y = 0$. Figure 18 shows the results after 6000 steps. The dots represent the locations of wake vortices, and the solid line is the path of the incident vortex. The path without wake vortices considered is plotted by the broken line for comparison. The path of the incident vortex with the wake vortices was rolled up to make a spiral through the interaction with the wake vortices and lingered in a region not far from the half-plane.

Figure 19 shows the acoustic pressure produced by the interaction of the incident vortex, wake vortices, and half-plane (solid line). The acoustic pressure without wake vortices is plotted by the broken line for comparison. According to Howe, the time is non-dimensionalized as UT/l, where U is $\Gamma/(8\pi l)$ and l is the distance of the closest incident vortex to the edge (where it crosses the x-axis at time 0 in Fig. 18) when no wake vortices are considered.

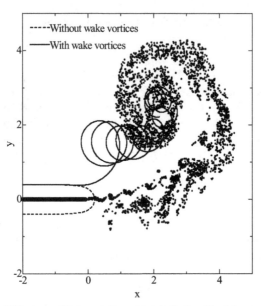

Fig. 18. Results after 6000 steps. Wake vortices (dots). Path of incident vortex with wake vortices considered (solid line). Path of incident vortex without wake vortices considered (broken line).

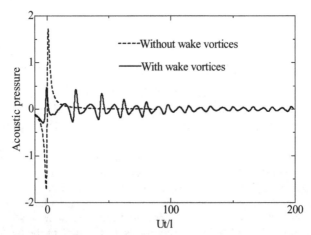

Fig. 19. Acoustic pressure. Calculated with wake vortices considered (solid line). Without wake vortices (broken line).

The peak values without wake vortices at around time = 0 were -1.72 (time = -0.790) and 1.72 (0.835), while the peak values with wake vortices were -0.29 (-3.28) and 0.47 (-0.449). This difference in values shows that the wake vortices decreased the peak values. However, the pressure continued to oscillate with the rolling movements of the incident vortex and wake vortices while it converged to zero when no wake vortices were considered.

Figure 20 shows the strength of vortices shed from the edge. When the incident vortex was far enough from the edge, wake vortices of negligible strength were produced. As the incident vortex approached the edge, the strength increased, and the wake with the largest strength of -0.0119 was produced at around time 0 when the incident vortex passed the edge. After this time, the strength of the wake vortices decreased, but oscillation due to the rolling movement was observed. The total strength of the wake vortices in Fig. 21 suddenly decreases at around time = 0, and it converged on the value of -0.9429, which is not 1 but the strength of the incident vortex. This vortex strength remains because the incident vortex did not flow away from the half-plane, and its effects on the production of wake vortices remained.

6. Three dimensions

Many factors in three-dimensional analyses affect the acoustic pressure. The effect of a change in the shape of an incident vortex on the pressure is presented here.

We considered a rectangular solid of dimensions $10 \times 10 \times 1$ with faces parallel to the x, y, and z axes with its center at the origin. An incident vortex of length 20 and strength 4π was placed from $x = -10$, $y = -10$ to $x = -10$, $y = 10$ with a height of $z = 0.5$ in a uniform flow of speed 1. The production of the wake vortices was not considered.

The rectangular solid was represented by distributions of source and vorticity, whose strength was determined by the non-permeable condition at each time step. The surfaces of the solid were divided into 2150 panels. The bound vortices Γ^{j}_{Bi} that represent the body in the unit flow in the j direction ($j = x$, y and z) were obtained in advance with the non-permeable condition imposed at the center of each panel. The time step was 0.005.

Figure 22(a) shows the locations for every 50 time steps of the incident vortex flowing from the left to the right in the uniform flow without changing its shape. The velocity at which the vortex moved was calculated at the center of the vortex. Figure 22(b) shows the incident vortex, which was divided into 20 elements of the same length so that the shape of the vortex could change according to the velocity induced at each element. The shape of the vortex deformed to fit the outline of the solid, and the vortex took almost twice as much time as that shown in Fig. 22(a) to flow from the leading edge of the solid to the trailing edge.

Figure 23 compares the acoustic pressures of the drag and lift dipoles, as normalized by $\rho_0 x_1 L U \Gamma / (4\pi c_0 |x|^2)$, of the rectangular solid with vortex deformation (broken line), the solid without vortex deformation (thick solid line), and, for reference, a two-dimensional flat plate from quasi three-dimensional simulation (thin solid line). Time t was normalized as Ut/a, where U is the uniform velocity of 1 and a is the half-length of the solid of 5.

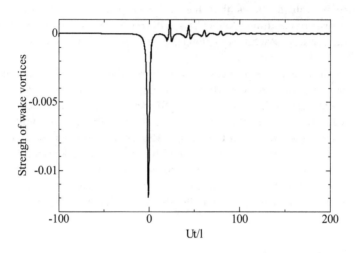

Fig. 20. Strength of wake vortices.

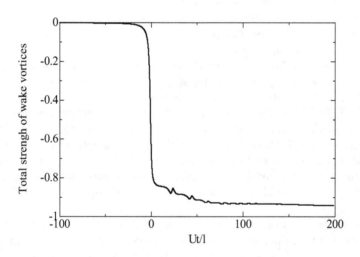

Fig. 21. Total strength of wake vortices.

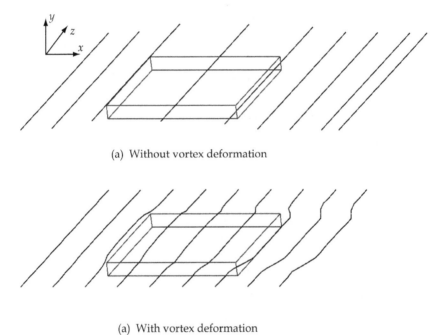

(a) Without vortex deformation

(a) With vortex deformation

Fig. 22. An incident vortex passing a rectangular solid with and without changing its shape.

(a)

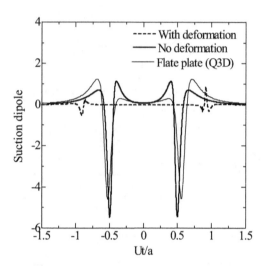

(b)

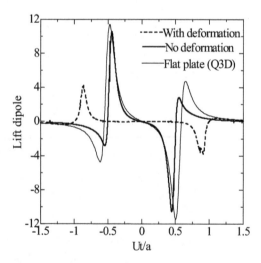

(a) Drag dipole; (b) Lift dipole

Fig. 23. Comparison of acoustic pressures for rectangular solid with vortex deformation (broken line), without vortex deformation (thick solid line), and a flat plate with quasi three-dimensional simulation (thin solid line).

The pressures of the flat plate and rectangular solid without vortex deformation were similar, while these for the rectangular solid with vortex deformation were rather weakened compared with line vortex analysis without a change in shape. The reason for this was as follows: In this study, the movement of an entire vortex line without deformation was calculated using the velocity at the center of the vortex, while the movement of a vortex line, divided into multiple elements, with deformation was estimated using the velocity at each element. Accordingly, as shown in Fig. 23(a), the time period between the negative peaks was almost 1 for "No deformation" and almost 2 for "With deformation." This shows that the time period for the incident vortex to travel from the leading edge to the trailing edge in the case "With deformation" was twice the value obtained in the case "No deformation." This implied that the velocity and acceleration of the incident vortex were evaluated to be larger for the "No deformation" case than for the "With deformation" case, resulting in a decrease in the pressure with an increase in the vortex deformation, as shown in Fig. 23.

7. Conclusion

We generalized Howe's method, based on the compact assumption that the wavelength of sound is large compared to the dimension of the solid body, to treat two-dimensional shapes with unknown conformal mappings and three-dimensional obstacles. The idea for calculating the acoustic pressure of arbitrarily shaped obstacles without conformal mappings is to use three sets of bound vortices which represent the obstacles in uniform flows of speed 1 in the x and y directions in two dimensions as well as in the z direction in three dimensions. These bound vortices are obtained by the panel method in advance. The acoustic pressures obtained by our method are considerably simple and easy to code.

The results of our method for a circular cylinder agreed well with Howe's solutions, and those for an NACA0012 airfoil agreed well with Kao's solutions through the asymptotic matching method.

Although Howe reported that the wake vortex shed from the trailing edge of an airfoil or half-plane cancels the acoustic pressure when the Kutta condition is employed, our simulations showed that (1) the production of the wake vortices decreases the acoustic pressure before the incident vortex reaches the edge, (2) the non-linear movements of the wake vortices keeps producing the acoustic pressure, and (3) the production of a large wake vortex enlarges the acoustic pressure. In addition, a change in the shape of the incident vortex decreases the acoustic pressure.

The simulations conducted here were intended to show the validity and accuracy of our method and to study very fundamental problems. Applications of our method to engineering problems will be shown in the near future. In addition, a vortex method for the vortex sound without the compact assumption is the next goal.

8. References

Curle, N. (1955). The influence of solid boundaries upon aerodynamic sound, *Proceedings of the Royal Society of London*, A231, pp. 505-514

Ffowcs Williams, J. E. & Hawkings, D. L. (1969). Sound generation by turbulence and surfaces in arbitrary motion, *Philosophical Transactions of the Royal Society of London*, A264, pp. 321-342

Howe, M. S. (1976). The influence of vortex shedding on the generation of sound by convected turbulence, *J. Fluid Mech.*, Vol. 76, Part 4, pp. 771-740

Howe, M. S. (2003). *Theory of Vortex Sound*, Cambridge University Press, ISBN 0-521-01223-6

Huberson, S.; Rivoalen, E. & Voutsinas, S. (2008). Vortex particle methods in aeroacoustic calculations, *J. Comp. Physics*, Vol. 227, pp. 9216-9240

Kao, H.C. (2002). Body-Vortex Interaction, Sound Generation, and Destructive Interface, *AIAA journal*, Vol. 40, No. 4, pp. 652-660.

Ogami, Y. & Akishita, S. (2010). Numerical simulation of sound field generated by separated flows using the vortex method, *Fluid Dynamics Research*, Vol. 42, No. 1, DOI 10.1088/0169-5983/42/1/015009

Powell, A. (1964). Theory of Vortex Sound, *Journal of the Acoustical Society of America*, Vol. 36, No. 1, pp. 177-195

Sharland, I.J. (1964). Sources of noise in axial flow fans, *J. Sound Vib.*, Vol. 1, No. 3, pp. 302-322

Part 4

Non-Classical Approaches

9

Topology Optimization of Fluid Mechanics Problems

Maatoug Hassine
ESSTH, Sousse University
Tunisia

1. Introduction

Optimal shape design problems in fluid mechanics have wide and valuable applications in aerodynamic and hydrodynamic problems such as the design of car hoods, airplane wings and inlet shapes for jet engines. One of the first studies is found in Pironneau (1974). It is devoted to determine a minimum drag profile submerged in a homogeneous, steady, viscous fluid by using optimal control theories for distributed parameter systems. Next, many shape optimization methods are introduced to determine the design of minimum drag bodies Kim and Kim (1995); Pironneau (1984), diffusers Cabuk and Modi (1992), valves Lund et al. (2002), and airfoils Cliff et al. (1998). The majority of works dealing with optimal design of flow domains fall into the category of shape optimization and are limited to determine the optimal shape of an existing boundary.

It is only recently that topological optimization has been developed and used in fluid design problems. It can be used to design features within the domain allowing new boundaries to be introduced into the design. In this context, Borvall and Petersson Borrvall and Petersson (2003) implemented the relaxed material distribution approach to minimize the power dissipated in Stokes flow. To approximate the no-slip condition along the solid-fluid interface they used a generalized Stokes problem to model fluid flow throughout the domain. Later, this approach was generalized by Guest and Prévast in Guest and Prévost (2006). They treated the material phase as a porous medium where fluid flow is governed by Darcy's law. For impermeable solid material, the no-slip condition is simulated by using a small value for the material permeability to obtain negligible fluid velocities at the nodes of solid elements. The flow regularization is expressed as a system of equations; Stokes flow governs in void elements and Darcy flow governs in solid elements.

In this work, we propose a new topological optimization method. Our approach is based on topological sensitivity analysis Amstutz (2005); Amstuts and Masmoudi (2003); Garreau et al. (2001); Guillaume and Hassine (2007); Guillaume and Sid Idris (2004); Hassine et al. (2007); Hassine and Masmoudi (2004); Masmoudi (2002); Sokolowski and Zochowski (1999). The optimal domain is constructed through the insertion of some obstacles in the initial one. The problem leads to optimize the obstacles location. The main idea is to compute the topological asymptotic expansion of a cost function j with respect to the insertion of a small obstacle inside the fluid flow domain. The obstacle is modeled as a small hole $\mathcal{O}_{z,\varepsilon}$ around a point z having an homogeneous condition on the boundary $\partial \mathcal{O}_{z,\varepsilon}$. The best location z of $\mathcal{O}_{z,\varepsilon}$ is given by the most negative value of a scalar function δj, called the topological gradient.

In practice, this approach leads to a simple, fast and accurate topological optimization algorithm. The final domain is obtained using an iterative process building a sequence of geometries $(\Omega_k)_k$ starting with the initial fluid flow domain $\Omega_0 = \Omega$. Knowing Ω_k, the new domain Ω_{k+1} is obtained by inserting an obstacle $\mathcal{O}_k$ in the domain Ω_k; $\Omega_{k+1} = \Omega_k \backslash \overline{\mathcal{O}_k}$. The location and the shape of $\mathcal{O}_k$ are defined by a level set curve of the topological gradient δj_k

$$\mathcal{O}_k = \{x \in \Omega_k, \text{ such that } \delta j_k(x) \leq c_k\},$$

where c_k is a scalar parameter used to control the size of the inserted obstacle. The function δj_k is the leading term of the variation $j(\Omega_k \backslash \overline{\mathcal{O}_{z,\varepsilon}}) - j(\Omega_k)$.

The chapter is organized as follows. In the next section, we present the topological optimization problem related to the Stokes system. The aim is to determine the fluid flow domain minimizing a given cost function. To solve this optimization problem we will use the topological sensitivity analysis method described in the Section 3. It consists in studying the variation of a cost function j with respect to a topology modification of the domain. The most simple way of modifying the topology consists in creating a small hole in the domain. In the case of structural shape optimization, creating a hole means simply removing some material. In the case of fluid dynamics where the domain represents the fluid, creating a hole means inserting a small obstacle $\mathcal{O}$. The topological sensitivity tools which have been developed by several authors Garreau et al. (2001); Schumacher (1995); Sokolowski and Zochowski (1999) allow to find the place where creating a small hole will bring the best improvement of the cost function. The main theoretical results are described in Sections 3.2 and 3.3. In section 3.2, we derive an asymptotic expansion for an arbitrary cost function with respect to the insertion of a small obstacle inside the fluid flow domain. In section 3.3, we derive an asymptotic expansion for two standard examples of cost functions.

As application of the proposed topological optimization method, we consider in Section 4 some engineering applications commonly found in the fluid mechanics literature. In Section 4.1, we present the optimization algorithm. In Section 4.2, we treat the shape optimization of pipes in a cavity. The aim is to determine the optimal shape of the pipes that connect the inlet to the outlets of the cavity minimizing the dissipated power in the fluid. The optimization of injectors location in an eutrophized lake is discussed in Section 4.3. Section 4.4 concerns the approximation of a wanted flow using a topological perturbation of the domain.

2. Topological optimization problem

Let Ω be a bounded domain of $\mathbb{R}^d$, $d = 2,3$ with smooth boundary Γ. We consider an incompressible fluid flow in Ω described by the Stokes equations. The velocity field u and the pressure p satisty the system

$$\begin{cases} -\nu \Delta u + \nabla p = F & \text{in } \Omega \\ \text{div } u = 0 & \text{in } \Omega \\ u = 0 & \text{on } \Gamma, \end{cases} \tag{1}$$

where ν denotes the kinematic viscosity of the fluid, F is a given body force per unit of mass (gravitational force).

The aim is to determine the optimal geometry of the fluid flow domain minimizing a given design function j:

$$\min_{D \in \mathcal{D}_{ad}} j(D), \text{ such that } |D| \leq V_{desired},$$

where j has the form
$$j(D) = J(u_D),$$
with u_D is the velocity field solution to the Stokes system in D and $\mathcal{D}_{ad}$ is a given set of admissible domains.

Here $|D|$ is the Lebesgue measure of D and $V_{desired}$ denotes the target volume (weight).

To solve this shape optimization problem we shall use the topological sensitivity analysis method. It consists in studying the variation of the objective function J with respect to a small topological perturbation of the domain Ω.

2.1 Stokes equations in the perturbed domain

We denote by $\Omega \backslash \overline{\mathcal{O}_\varepsilon}$ the perturbed domain, obtained by inserting a small obstacle $\mathcal{O}_\varepsilon$ in Ω. We suppose that the obstacle has the form $\mathcal{O}_\varepsilon = x_0 + \varepsilon \mathcal{O}$, where $x_0 \in \Omega$, $\varepsilon > 0$ and $\mathcal{O}$ is a given fixed and bounded domain of $\mathbb{R}^d$, containing the origin, whose boundary $\partial \mathcal{O}$ is connected and piecewise of class $\mathcal{C}^1$.

In $\Omega \backslash \overline{\mathcal{O}_\varepsilon}$, the velocity u_ε and the pressure p_ε are solution to

$$\begin{cases} -\nu \Delta u_\varepsilon + \nabla p_\varepsilon = F & \text{in } \Omega \backslash \overline{\mathcal{O}_\varepsilon} \\ \operatorname{div} u_\varepsilon = 0 & \text{in } \Omega \backslash \overline{\mathcal{O}_\varepsilon} \\ u_\varepsilon = 0 & \text{on } \Gamma \\ u_\varepsilon = 0 & \text{on } \partial \mathcal{O}_\varepsilon. \end{cases} \tag{2}$$

Note that for $\varepsilon = 0$, $\Omega_0 = \Omega$ and (u_0, p_0) is solution to

$$\begin{cases} -\nu \Delta u_0 + \nabla p_0 = F & \text{in } \Omega, \\ \operatorname{div} u_0 = 0 & \text{in } \Omega, \\ u_0 = 0 & \text{on } \Gamma. \end{cases} \tag{3}$$

2.2 Topological optimization problem

Consider now a design function j of the form

$$j(\Omega \backslash \overline{\mathcal{O}_\varepsilon}) = J_\varepsilon(u_\varepsilon), \tag{4}$$

where J_ε is a given cost function defined on $H^1(\Omega \backslash \overline{\mathcal{O}_\varepsilon})^d$ for $\varepsilon \geq 0$ and u_ε is the velocity field solution to the Stokes system (2).

Our aim is to determine the optimal location of the obstacle $\mathcal{O}_\varepsilon$ in the domain Ω in order to minimize the design function j. Then, the optimization problem we consider is given as follows:

$$\min_{\mathcal{O}_\varepsilon \subset \Omega} j(\Omega \backslash \overline{\mathcal{O}_\varepsilon}). \tag{5}$$

To this end, we will derive in the next section a topological asymptotic expansion of the function j with respect to ε.

3. Topological sensitivity analysis

In this section we consider a topological sensitivity analysis for the Stokes equations. We present a topological asymptotic expansion of a design function j with respect to the insertion of a small obstacle $\mathcal{O}_\varepsilon$ inside the domain Ω. The proposed approach is based on the following general adjoint method.

3.1 General adjoint method

Let $(\mathcal{V}_\varepsilon)_{\varepsilon \geq 0}$ be a family of Hilbert spaces depending on the parameter ε, such that, $\forall \varepsilon \geq 0$ $\mathcal{V}_\varepsilon \hookrightarrow \mathcal{V}_0$. For $\varepsilon \geq 0$, we consider

- $\mathcal{A}_\varepsilon : \mathcal{V}_\varepsilon \times \mathcal{V}_\varepsilon \longrightarrow \mathbb{R}$ a bilinear, continuous and coercive form on $\mathcal{V}_\varepsilon$,
- $l_\varepsilon : \mathcal{V}_\varepsilon \longrightarrow \mathbb{R}$ a linear and continuous form on $\mathcal{V}_\varepsilon$.

For all $\varepsilon \geq 0$, we denote by u_ε the unique solution to the problem

$$\mathcal{A}_\varepsilon(u_\varepsilon, w) = l_\varepsilon(w), \quad \forall w \in \mathcal{V}_\varepsilon. \tag{6}$$

Consider now a cost function of the form $j(\varepsilon) = J_\varepsilon(u_\varepsilon)$, where J_ε is defined on $\mathcal{V}_\varepsilon$ for $\varepsilon \geq 0$ and J_0 is differentiable with respect to u, its derivative being denoted by $DJ_0(u)$.

Our aim is to derive an asymptotic expansion of j with respect to ε. We consider the following assumptions.

Hypothesis 3.1. *There exist a real number $\delta\mathcal{A}$ and a scalar function $f : \mathbb{R}_+ \longrightarrow \mathbb{R}_+$ such that $\forall \varepsilon \geq 0$*

$$\mathcal{A}_0(u_0 - u_\varepsilon, v_0) = f(\varepsilon)\delta\mathcal{A} + o(f(\varepsilon)),$$
$$\lim_{\varepsilon \to 0} f(\varepsilon) = 0,$$

where $v_0 \in \mathcal{V}_0$ is the solution to the adjoint problem

$$\mathcal{A}_0(w, v_0) = -DJ_0(u_0)w, \quad \forall w \in \mathcal{V}_0. \tag{7}$$

Hypothesis 3.2. *There exists a real number δJ such that $\forall \varepsilon \geq 0$*

$$J_\varepsilon(u_\varepsilon) - J_0(u_0) = DJ_0(u_0)(u_\varepsilon - u_0) + f(\varepsilon)\delta J + o(f(\varepsilon)).$$

Under the assumptions 3.1 and 3.2, we have the following theorem.

Theorem 3.1. *Hassine et al. (2008) If the assumptions 3.1 and 3.2 hold, the function j has the following asymptotic expansion*

$$j(\varepsilon) = j(0) + f(\varepsilon)\Big(\delta\mathcal{A} + \delta J\Big) + o(f(\varepsilon)).$$

3.2 Topological sensitivity for the Stokes problem

In this section, we derive a topological asymptotic expansion for the Stokes equations. In order to apply the adjoint method described in the previous paragraph, first we establish a variational problem associated to the Stokes system. From the weak variational formulation of (2), we deduce that $u_\varepsilon \in \mathcal{V}_\varepsilon$ is solution to

$$\mathcal{A}_\varepsilon(u_\varepsilon, w) = l_\varepsilon(w), \quad \forall w \in \mathcal{V}_\varepsilon,$$

where the functional space $\mathcal{V}_\varepsilon$, the bilinear form $\mathcal{A}_\varepsilon$ and the linear form l_ε are defined by

$$\mathcal{V}_\varepsilon = \left\{w \in H_0^1(\Omega_\varepsilon), \text{ div } w = 0 \text{ in } \Omega_\varepsilon\right\}, \tag{8}$$

$$\mathcal{A}_\varepsilon(v, w) = v \int_{\Omega_\varepsilon} \nabla v \cdot \nabla w \, dx, \quad \forall u, v \in \mathcal{V}_\varepsilon, \tag{9}$$

$$l_\varepsilon(w) = \int_{\Omega_\varepsilon} F w \, dx, \quad \forall w \in \mathcal{V}_\varepsilon, \tag{10}$$

where $\Omega_\varepsilon = \Omega \backslash \overline{\mathcal{O}_\varepsilon}$.

Next we have to distinguish the cases $d = 2$ and $d = 3$, because the fundamental solutions to the Stokes equations in $\mathbb{R}^2$ and $\mathbb{R}^3$ have an essentially different asymptotic behaviour at infinity.

3.2.1 The three dimensional case

Let (U, P) denote a solution to

$$\begin{cases} -\nu \Delta U + \nabla P = 0 & \text{in } \mathbb{R}^3 \backslash \overline{\mathcal{O}} \\ \operatorname{div} U = 0 & \text{in } \mathbb{R}^3 \backslash \overline{\mathcal{O}} \\ U \longrightarrow 0 & \text{at } \infty \\ U = -u_0(x_0) & \text{on } \partial\mathcal{O}. \end{cases} \tag{11}$$

The existence of (U, P) is most easily established by representing it as a single layer potential on $\partial\mathcal{O}$ (see Dautray and Lions (1987))

$$U(y) = \int_{\partial\mathcal{O}} E(y - x)\eta(x)\,\mathrm{d}s(x), \quad P(y) = \int_{\partial\mathcal{O}} \Pi(y - x)\eta(x)\,\mathrm{d}s(x), \quad y \in \mathbb{R}^3 \backslash \overline{\mathcal{O}}$$

where (E, Π) is the fundamental solution of the Stokes equations

$$E(y) = \frac{1}{8\pi\nu r}\left(I + e_r e_r^T\right), \quad \Pi(y) = \frac{y}{4\pi r^3},$$

with $r = \|y\|$, $e_r = y/r$ and e_r^T is the transposed vector of e_r. The function $\eta \in H^{-1/2}(\partial\mathcal{O})^3$ is the solution to the boundary integral equation,

$$\int_{\partial\mathcal{O}} E(y - x)\,\eta(x)\,\mathrm{d}s(x) = -u_0(x_0), \quad \forall y \in \partial\mathcal{O}. \tag{12}$$

One can observe that the function η is determined up to a function proportional to the normal, hence it is unique in $H^{-1/2}(\partial\mathcal{O})^3/\mathbb{R}n$.

We start the derivation of the topological asymptotic expansion with the following estimate of the $H^1(\Omega_\varepsilon)$ norm of $u_\varepsilon(x) - u_0(x) - U(x/\varepsilon)$. This estimate plays a crucial role in the derivation of our topological asymptotic expansion. It describes the velocity perturbation caused by the presence of the small obstacle $\mathcal{O}_\varepsilon$.

Proposition 3.1. *Guillaume and Hassine (2007); Hassine et al. (2008) There exists $c > 0$, independent on ε, such that for all $\varepsilon > 0$ we have*

$$\|u_\varepsilon(x) - u_0(x) - U(x/\varepsilon)\|_{1,\Omega_\varepsilon} \leq c\varepsilon.$$

The following corollary follows from Proposition 3.1. It gives the behaviour of the velocity u_ε when inserting an obstacle. The principal term of this perturbation is given by the function U, solution to (11).

Corollary 3.1. *We have*

$$u_\varepsilon(x) = u_0(x) + U(x/\varepsilon) + O(\varepsilon), \quad x \in \Omega_\varepsilon.$$

We are now ready to derive the topological asymptotic expansion of the cost function j. It consists in computing the variation $j(\Omega \backslash \overline{\mathcal{O}_\varepsilon}) - j(\Omega)$ when inserting a small obstacle inside the domain. The leading term of this variation involves the function η, the solution to the boundary integral equation (12). The main result is described by Theorem 3.2.

Theorem 3.2. *Guillaume and Hassine (2007); Hassine et al. (2008) If Hypothesis 3.1 holds, the function j has the following asymptotic expansion*

$$j(\Omega \backslash \overline{\mathcal{O}_\varepsilon}) = j(\Omega) + \varepsilon \, \delta j(x_0) + o(\varepsilon).$$

where the topological gradient δj is given by

$$\delta j(x) = \left(- \int_{\partial \mathcal{O}} \eta(y) \, ds(y) \right) \cdot v_0(x) + \delta J(x), \quad x \in \Omega.$$

If $\mathcal{O}$ is the unit ball centred at the origin, $\mathcal{O} = B(0,1)$, the density η is given explicitly $\eta(y) = -\dfrac{3v}{2} u_0(x_0), \forall y \in \partial \mathcal{O}.$

Corollary 3.2. *If $\mathcal{O} = B(0,1)$, under the hypotheses of theorem 3.2, we have*

$$j(\Omega \backslash \overline{\mathcal{O}_\varepsilon}) = j(\Omega) + \varepsilon \left[6\pi v \, u_0(x_0) \cdot v_0(x_0) + \delta J(x_0) \right] + o(\varepsilon).$$

3.2.2 The two dimensional case

In this paragraph, we present the topological asymptotic expansion for the Stokes equations in the two dimensional case. The result is obtained using the same technique described in the previous paragraph. The unique difference comes from the expression of the fundamental solution of the Stokes equations. In this case (E, Π) is given by

$$E(y) = \frac{1}{4\pi v} \left(-\log(r)I + e_r e_r^T \right), \quad \Pi(y) = \frac{y}{2\pi r^2}.$$

Theorem 3.3. *Guillaume and Hassine (2007); Hassine et al. (2008) Under the same hypotheses of theorem 3.2, the function j has the following asymptotic expansion*

$$j(\Omega \backslash \overline{\mathcal{O}_\varepsilon}) = j(\Omega) + \frac{-1}{\log(\varepsilon)} \, \delta j(x_0) + o\left(\frac{-1}{\log(\varepsilon)} \right).$$

where the topological gradient δj is given by

$$\delta j(x) = 4\pi v \, u_0(x) \cdot v_0(x) + \delta J(x), \quad x \in \Omega.$$

3.3 Cost function examples

We now discuss Assumption 3.2. We present two standard examples of cost functions satisfying this Assumption and we calculate their variations δJ. For the proofs one can see Guillaume and Hassine (2007) or Hassine et al. (2008).

Proposition 3.2. *Let $w_d \in H^1(\Omega)$ be a given wanted (objective) velocity field. The cost function*

$$J_\varepsilon(u) = \int_{\Omega \backslash \overline{\mathcal{O}_\varepsilon}} |u - w_d|^2 \, dx, \tag{13}$$

satisfies the assumption 3.1 with

$$DJ_0(w) = 2 \int_{\Omega} (u_0 - w_d) \cdot w \, dx, \quad \forall w \in \mathcal{V}_0, \text{ and } \delta J(x_0) = 0.$$

Proposition 3.3. *Let $w_d \in H^2(\Omega)$. The cost function*

$$J_\varepsilon(u) = \nu \int_{\Omega \setminus \overline{\mathcal{O}_\varepsilon}} |\nabla u - \nabla w_d|^2 \, dx, \tag{14}$$

satisfies the assumption 3.1 with

$$DJ_0(w) = 2 \int_\Omega \nabla(u_0 - w_d) \cdot \nabla w \, dx \quad \forall w \in V_0,$$

$$\delta J(x_0) = \begin{cases} \left(-\int_{\partial \mathcal{O}} \eta(y) \, ds(y) \right) \cdot u_0(x_0) & \text{if } d = 3, \\ 4\pi \nu |u_0(x_0)|^2 & \text{if } d = 2. \end{cases}$$

For d=3, if $\mathcal{O}$ is the unit ball $B(0,1)$, we have $\delta J = 6\pi \nu |u_0(x_0)|^2$.

4. Numerical experiments

As an application of the previous theoretical results, we consider some engineering applications commonly found in the fluid mechanics literature. Our implementation is based on the following optimization algorithm.

4.1 The optimization algorithm

We apply an iterative process to build a sequence of geometries $(\Omega_k)_{k \geq 0}$ with $\Omega_0 = \Omega$. At the kth iteration the topological gradient is denoted by δj_k and the new geometry Ω_{k+1} is obtained by inserting an obstacle $\mathcal{O}_k$ in the domain Ω_k; $\Omega_{k+1} = \Omega_k \setminus \overline{\mathcal{O}_k}$. The location and the size of the obstacle $\mathcal{O}_k$ are chosen in such a way that $j(\Omega_{k+1}) - j(\Omega_k)$ is negative.

Based on the last remark, the obstacle $\mathcal{O}_k$ is defined by a level set curve of the topological gradient δj_k

$$\mathcal{O}_k = \left\{ x \in \Omega_k, \text{ such that } \delta j_k(x) \leq c_k \leq 0 \right\},$$

where c_k is chosen in such a way that $|\mathcal{O}_k|/|\Omega_k|$ is less than a given ratio $\delta \in]0, 1[$.

The algorithm : Topology optimization with volume constraint.
- Initialization: choose $\Omega_0 = \Omega$, and set $k = 0$.
- Repeat until $|\Omega_k| \leq V_{desired}$:

- compute u_k the solution to the Stokes equations (15) in Ω_k,

- compute v_k the solution to the associated adjoint problem (16) in Ω_k ,

- compute the topological sensitivity $\delta j_k(z)$, $\forall z \in \Omega_k$,

- determine $\Omega_{k+1} = \Omega_k \setminus \overline{\mathcal{O}_k}$, where $\mathcal{O}_k = \left\{ x \in \Omega_k, \text{ such that } \delta j_k(x) \leq c_k \leq 0 \right\}$,

- $k \longleftarrow k + 1$.

The topological gradient δj_k is defined by

$$\delta j_k(z) = u_k(z) \cdot v_k(z) + \delta J_k(z), \quad \forall z \in \Omega_k,$$

where u_k is the velocity field solution to

$$\begin{cases} -\nu \Delta u_k + \nabla p_k = F & \text{in } \Omega_k \\ \text{div } u_k = 0 & \text{in } \Omega_k, \end{cases} \tag{15}$$

and v_k is the solution to the associated adjoint problem

$$\begin{cases} -\nu \, \Delta v_k + \nabla q_k = -DJ(u_k) & \text{in } \Omega_k \\ \qquad\quad \text{div } v_k = 0 & \text{in } \Omega_k. \end{cases} \tag{16}$$

The discretization of the problems (15) and (16) is based on the mixed finite element method $P1 + bubble/P1$ Arnold et al. (1984). The function δj_k is computed piecewise constant over elements. The term δJ_k is the variation of the considered cost function J (see Propositions 3.2 and 3.3). The constant c_k determines the volume of the obstacle $\mathcal{O}_k$ to be inserted. In practice, c_k is chosen in such a way that:

i- $\mathcal{O}_k \subset \left\{ x \in \Omega_k, \text{ such that } \delta j_k(x) \leq 0 \right\}$,

ii- the obstacle volume $|\mathcal{O}_k|$ is less or equal to 10% of the current domain volume $|\Omega_k|$ i.e. $|\mathcal{O}_k|/|\Omega_k| \leq 0.1$.

This algorithm can be seen as a descent method where the descent direction is determined by the topological sensitivity δj_k and the step length is given by the volume variation $|\Omega_k \backslash \Omega_{k+1}|$.

4.2 Pipes shape optimization

We consider a viscous and incompressible fluid in a tank Ω having one inlet Γ_{in} and some outlets Γ_{out}^i, $1 \leq i \leq m$. The aim is to determine the optimal design of the pipes that connects the inlet to the outlet of the domain minimizing the dissipated power in the fluid.

4.2.1 Comparison

In order to test the advantage of our approach, we compare our results to those obtained in Borrvall and Petersson (2003); Glowinski and Pironneau (1975). We consider two numerical examples in two dimensional (2D) case. The first one is the pipe bend example presented in Figure 1. This test case is treated by Borrvall and Petersson in Borrvall and Petersson (2003). The second one is the double pipe shown in Figure 2. It is also considered by Borrvall and Petersson in Borrvall and Petersson (2003) and recently by Guest and Prévost in Glowinski and Pironneau (1975). The aim here is to obtain the optimal shape minimizing the dissipated power in the fluid.

The considered design function is given by

$$j(D) = \nu \int_D |\nabla u_D|^2 \, dx,$$

where u_D is the solution to the Stokes system in D.

The optimization problem consists in finding the fluid flow domain solution to

$$\min_{D \in \mathcal{D}_{ad}} j(D), \text{ such that } |D| \leq V_{desired}$$

where $\mathcal{D}_{ad}$ is the set of admissible domains defined by

$$\mathcal{D}_{ad} = \{D \subset \Omega \text{ such that } \Gamma_{in} \subset \partial\Omega \cap \partial D \text{ and } \Gamma_{out}^i \subset \partial\Omega \cap \partial D\}.$$

In both cases the inflow and the outflow conditions are given by a parabolic flow profile type with a maximum flow velocity equal to 1. Elsewhere the velocity is prescribed to be zero on the boundary of the domain.

A- Test 1 : *2D pipe bend example.* We consider a cavity $\Omega =]0, 1[\times]0, 1[$ having one inlet (left) and one outlet (bottom) (see figure 1(a)).

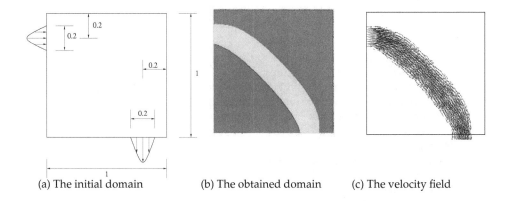

(a) The initial domain   (b) The obtained domain   (c) The velocity field

Fig. 1. 2D pipe bend example.

The cavity Ω is discretized using a finite elements mesh with 6561 nodes and 12800 triangular elements. The results of this example are presented in figure 1. The obtained pipe geometry is described in figure 1(b). It is computed using $V_{desired} = 0.08\pi \, |\Omega|$. The prescribed volume constraint is chosen so that the optimal solution has the same volume as a quarter torus of inner radius 0.7 and outer radius 0.9 that exactly fits to the inlet and outlet. In figure 1(c) we present the velocity field computed in the final domain.

The obtained solution is nearly identical to those presented in Borrvall and Petersson Borrvall and Petersson (2003). However, we obtain this result in 14 iterations, where Borrvall and Petersson needed more than sixty. As it can be seen, we have a more torus shaped pipe than in Borrvall and Petersson (2003), like most pipe bends in fluid mechanics literature. As it is stated in Glowinski and Pironneau (1975), the solution in Borrvall and Petersson Borrvall and Petersson (2003) contains regions of artificial material and does not sufficiently take into account the adherence condition.

B- Test 2: *2D double pipe example.* The initial domain of this example is shown in Figure 2(a). It is the rectangular $\Omega =]0, 3/2[\times]0, 1[$ with two inlets and two outlets.

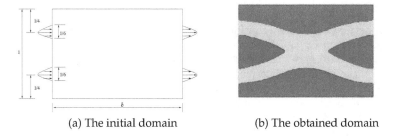

(a) The initial domain   (b) The obtained domain

Fig. 2. The initial and the optimal domains for the 2D double pipe example.

The cavity Ω is discretized using a finite elements mesh with 9801 nodes and 19200 triangular elements. The results of this example are presented in figure 2. The final geometry is computed with $V_{desired} = \dfrac{1}{3} \, |\Omega|$.

We present in figure 2(b) the obtained geometry. The final geometry is obtained in only 12 iterations, where Borrvall and Petersson needed more than sixty. We remark that the two pipes join to form a single, wider pipe through the center of the domain. This design decreases the length of the fluid-solid interface by decreasing the power lost. As it can be seen, the optimal solution is identical to that obtained by Guest and Prévost Glowinski and Pironneau (1975), but it does not match that of Borrvall and Petersson Borrvall and Petersson (2003). As for the pipe bend example, the solution in Borrvall and Petersson (2003) contains regions of artificial material and does not sufficiently take into account the adherence condition.

4.2.2 Three dimensional case
In this section we propose an extension of the two 2D examples considered in the last section to the three dimensional case.

A- Example 1 : *3D pipe bend example.* For the 3D pipe bend example, the initial domain is the unit cube $\Omega =]0,1[\times]0,1[\times]0,1[$ having one inlet and one outlet (see figure 3). The inlet Γ_{in} (left) and the outlet Γ_{out} (bottom) are described by the following discs

$$\Gamma_{in} = B(z_{in},0.1) \cap \{0\}\times]0,1[\times]0,1[, \text{ and } \Gamma_{out} = B(z_{out},0.1)\cap]0,1[\times]0,1[\times\{0\},$$

where $B(z_\beta,0.1)$, $\beta = in, out$, is the ball of center z and radius 0.1, with $z_{in} = (0,0.5,0.8)$ and $z_{out} = (0.8,0.5,0)$.

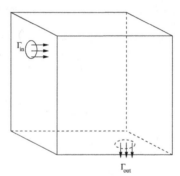

Fig. 3. The initial domain

For the boundary conditions, we consider a parabolic flow profile type with a maximum flow velocity equal to 1 on Γ_{in} and Γ_{out}, and a velocity equal to zero elsewhere. The domain is discretized using 29791 nodes and 162000 tetrahedral elements.

Like in the 2D case, we aim to determine the optimal design of the pipe that connects the inlet to the outlet of the domain and minimizes the dissipated power in the fluid. We present in figure 4 the optimal pipe domains obtained for different volume constraint $V_{desired}$ choices. The first case (figure 4(a)), corresponding to $V_{desired} = 0.50\,|\Omega|$, is obtained after 7 iterations, the second one (figure 4(b)) after 11 iterations for $V_{desired} = 0.35\,|\Omega|$ and the last one (figure 4(b)) needs 16 iterations to reach $V_{desired} = 0.20\,|\Omega|$. We show in figure 5 a 2D cut of the velocity field corresponding to the three obtained domains.

B- Example 2 : *3D double pipe bend example.* The initial domain is the cavity $\Omega =]0,3/2[\times]0,1[\times]0,1[$ (described in figure 6). It has two inlets (left) Γ_{in}^i, i=1,2, and two outlets

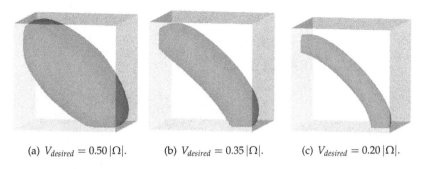

(a) $V_{desired} = 0.50\,|\Omega|$.   (b) $V_{desired} = 0.35\,|\Omega|$.   (c) $V_{desired} = 0.20\,|\Omega|$.

Fig. 4. The obtained domains (see Abdelwahed, Hassine and Masmoudi (2009)).

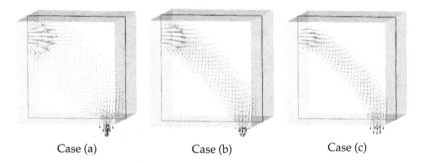

Case (a)   Case (b)   Case (c)

Fig. 5. 2D vertical cut of the velocity field in the obtained domains.

(right) Γ_{out}^i, i=1,2 defined by

$$\Gamma_{in}^1 = B(z_{in}^1, 0.1) \cap \{0\} \times]0, 1[\times]0, 1[, \ \Gamma_{in}^2 = B(z_{in}^2, 0.1) \cap \{0\} \times]0, 1[\times]0, 1[,$$
$$\Gamma_{out}^1 = B(z_{out}^1, 0.1) \cap \{3/2\} \times]0, 1[\times]0, 1[, \ \Gamma_{out}^2 = B(z_{out}^2, 0.1) \cap \{3/2\} \times]0, 1[\times]0, 1[,$$

where

$$z_{in}^1 = (0, 1/2, 1/4), \ z_{in}^2 = (0, 1/2, 3/4), \ z_{out}^1 = (3/2, 1/2, 1/4), \text{ and } z_{out}^2 = (3/2, 1/2, 3/4)$$

For the boundary conditions, as in the last example, we consider a parabolic flow profile type with a maximum flow velocity equal to 1 on Γ_{in}^i and on Γ_{out}^i, and a velocity equal to zero elsewhere. We use a mesh with 160602 nodes and 895900 tetrahedral elements.

We present in figure 7 the optimal shape design obtained respectively for $V_{desired} = 0.40\,|\Omega|$ (9 iterations) and $V_{desired} = 0.10\,|\Omega|$ (21 iterations). A vertical cut of the corresponding velocity field is shown in figure 8.

4.2.3 Shape optimization of tubes in a 3D cavity

In this section we treat the shape optimization of tubes in a cavity. We consider an incompressible fluid in a cavity Ω having one inlet Γ_{in} and four outlets Γ_{out}^i, $i = 1, 4$. The aim here is to determine the optimal shape of the tubes that connect the inlet to the outlets of the cavity maximizing the outflow rate. It consists in inserting small obstacles in the cavity in order to maximize the outflow rate at Γ_{out}^i, $i = 1, 4$.

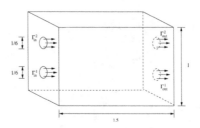

Fig. 6. The initial domain

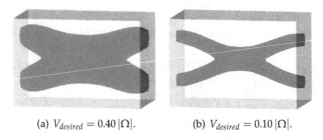

(a) $V_{desired} = 0.40\,|\Omega|.$ (b) $V_{desired} = 0.10\,|\Omega|.$

Fig. 7. The optimal domains (see Abdelwahed, Hassine and Masmoudi (2009)).

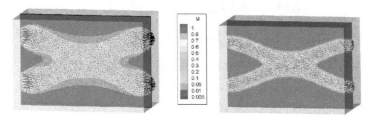

Fig. 8. 2D vertical cut of the velocity isovalues and field in the optimal domains.

In our numerical computation, we have used the cavity $\Omega =]0,1[\times]0,1[\times]0,1[$ with the inlet Γ_{in}:

$$\Gamma_{in} = \left\{(x,y,z) \in \Omega \text{ such that } x^2 + (y - 0.5)^2 + (z - 0.5)^2 \le 0.04\right\}$$

and the four outlets $\Gamma^1_{out}, \Gamma^2_{out}, \Gamma^3_{out}$ and Γ^4_{out}:

$$\Gamma^1_{out} = \left\{(x,y,z) \in \Omega \text{ such that } (x - 0.75)^2 + (y - 0.5)^2 + z^2 \le 0.0025\right\},$$

$$\Gamma^2_{out} = \left\{(x,y,z) \in \Omega \text{ such that } (x - 0.75)^2 + (y - 0.5)^2 + (z - 1)^2 \le 0.0025\right\},$$

$$\Gamma^3_{out} = \left\{(x,y,z) \in \Omega \text{ such that } (x - 0.75)^2 + y^2 + (z - 0.5)^2 \le 0.0025\right\},$$

$$\Gamma^4_{out} = \left\{(x,y,z) \in \Omega \text{ such that } (x - 0.75)^2 + (y - 1)^2 + (z - 0.5)^2 \le 0.0025\right\}.$$

The considered cost function measuring the outflow rate is given by

$$j(D) = \sum_{i=1}^{m} \int_{\Gamma^i_{out}} |u_D.n|\, ds,$$

where $D \in \mathcal{D}_{ad}$ and u_D is the velocity field, solution to the Stokes equations in D satisfying the following boundary conditions :

- A free surface boundary condition on the outlets

$$\sigma(u) \cdot \mathbf{n} = 0 \text{ on } \cup_{i=1}^{m} \Gamma_{out}^{i},$$

where $\sigma(u) = \nu(\nabla u + \nabla u^{T}) - pI$, I is the 3×3 identity matrix and $\mathbf{n}$ denotes the outward normal to the boundary.

- The normal component of the stress tensor is prescribed on the inlet Γ_{in}

$$\sigma(u) \cdot \mathbf{n} = g \text{ on } \Gamma_{in},$$

- The velocity is equal to zero on $\Gamma \backslash (\cup_{i=1}^{m} \Gamma_{out}^{i} \cup \Gamma_{in})$.

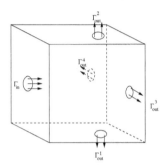

Fig. 9. The initial domain.

The results of this example are described in figures 10-13. In figure 10 we present the obtained geometries for different volume constraints. We present the obtained geometry: in figure10(a) for $V_{desired} = 0.35 |\Omega|$, in figure10(b) for $V_{desired} = 0.25 |\Omega|$ and in figure10(b) for $V_{desired} = 0.15 |\Omega|$. This domains are obtained respectively after 10, 14 and 19 iterations. The associated velocities fields are given in figures 11 and 12. In figure 13 we illustrate the variation of the outflow rate.

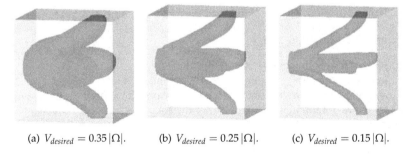

(a) $V_{desired} = 0.35 |\Omega|$.   (b) $V_{desired} = 0.25 |\Omega|$.   (c) $V_{desired} = 0.15 |\Omega|$.

Fig. 10. The optimal domains (see Abdelwahed, Hassine and Masmoudi (2009)).

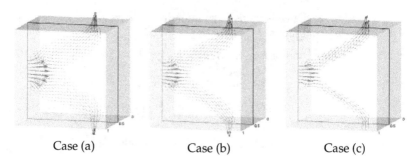

Case (a)   Case (b)   Case (c)

Fig. 11. 2D vertical cut of the velocity field.

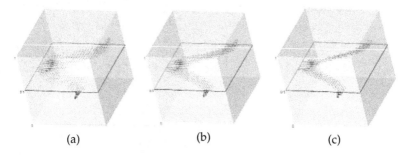

(a)   (b)   (c)

Fig. 12. 2D horizontal cut of the velocity field.

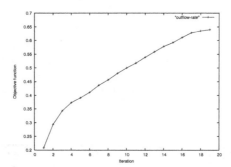

Fig. 13. Variation of the outflow rate

4.3 Optimization of injectors location in an eutrophic lake

Eutrophication is a complex phenomena involving many physico-chemical parameters. Specifically in some climatic areas, the thermic factors combined to the biological and to the biochemical ones are dominant in the behavior of the aquatic ecosystems. Consequently, they generate important bio-climatology variations creating in lakes an unsteady dynamic process that decreases progressively water quality. Practically, the eutrophication in a water basin is characterized mainly by a poor dissolved oxygen concentration in water. Furthermore, this phenomena is accompanied by a stratification process dividing the water volume, during a large period of the year, into three distinct layers as depicted in Figure 14. Three zones constitute this stratification:

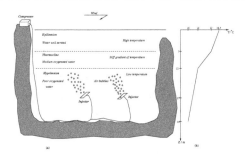

Fig. 14. (a): Structure of a stratified lake, (b): average temperature curve during summer

i) at the top, the epilimnion, a layer of around 7 m depth, well mixed by the effect of drafting wind and consequently well aerated,

ii) in the middle, the thermocline, a zone with a quick decrease of temperature (27 °C to 18 °C) and of 5 m depth. This area is weakly affected by the wind action and consequently a medium concentration of oxygen is observed,

iii) at the bottom, the hypolimnion, a deeper layer beyond 12 m, having a temperature varying from 18 °C to 14 °C. This region is characterized by a low concentration of oxygen and a high concentration of toxic gas (H_2S, ammoniac, carbonic gas, etc.)

The dynamic aeration process seems to be the most promising remedial technique to treat water eutrophication. This technique consists in inserting air by the means of injectors located at the bottom of the lake in order to generate a vertical motion mixing up the water of the bottom with that in the top, thus oxygenating the lower part by bringing it in contact with the surface air.

Theoretically, the bubble flow is a multi-phase flow where the presence of free interfaces raises difficulties in both the physical and mathematical modelling. Hence, to obtain a physical and significant resolution by numerical simulation of the air injection phenomena in an eutrophised lake, one should consider a two-phase model: water-air bubble (see Ishii Ishii (1975)). This kind of modelling involves large systems of PDE's and variables in a multi-scale frame as well as closure conditions through turbulence model and phases interface interaction. Moreover, the domain mesh size should be "small" in order to capture the significant variations of the spectrum. Therefore, the computational cost should be also addressed.

For all these reasons, we consider here, as a first approximation, only the liquid phase, which is the dominant one. The flow is described by a simplified model based on incompressible Stokes equations. The injected air is taken into account through local boundary conditions for the velocity on the injectors holes. In order to generate the best motion in the fluid with respect to the aeration purpose, the topological sensitivity analysis method is used to optimize the injectors location.

4.3.1 Optimization problem

In this section, we use the topological sensitivity analysis method to optimize the injector locations in the lake Ω in order to generate the best motion in the fluid with respect to the aeration purpose.

To this end, each injector Inj_k is modeled as a small hole $\mathcal{B}_{z_k,\varepsilon} = z_k + \varepsilon\mathcal{B}^k$, $1 \leq k \leq m$ having an injection velocity u_{inj}^k, where ε is the shared diameter and $\mathcal{B}^k \subset \mathbb{R}^d$ are bounded and smooth domains containing the origin. The points $z_k \in \Omega$, $1 \leq k \leq m$ determine the location of the injectors.

Then, in the presence of injectors, the velocity u_ε and the pressure p_ε satisfy the following system

$$
\begin{cases}
-\nu \Delta u_\varepsilon + \nabla p_\varepsilon = F & \text{in } \Omega \backslash \cup_{k=1}^{m} \overline{\mathcal{B}_{z_k,\varepsilon}} \\
\operatorname{div} u_\varepsilon = 0 & \text{in } \Omega \backslash \cup_{k=1}^{m} \overline{\mathcal{B}_{z_k,\varepsilon}} \\
u_\varepsilon = u_d & \text{on } \Gamma \\
u_\varepsilon = u_{inj}^k & \text{on } \cup_{k=1}^{m} \partial \mathcal{B}_{z_k,\varepsilon},
\end{cases}
\tag{17}
$$

where u_{inj}^k is a given injection velocity on $\partial \mathcal{B}_{z_k,\varepsilon}$, $1 \leq k \leq m$.

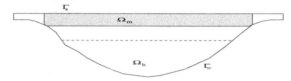

Fig. 15. The geometry of the lake.

Concerning the optimization criteria, we assume that a "good" lake oxygenation can be described by a target velocity U_g. Then, the cost function J_ε to be minimized is defined by

$$
J_\varepsilon(u_\varepsilon) = \int_{\Omega_m} |u_\varepsilon - U_g|^2 \, dx,
\tag{18}
$$

where $\Omega_m \subset \Omega$ is the measurement domain (the top layer, see Figure 15).
Consider the design function j of the form

$$
j(\Omega \backslash \cup_{k=1}^{m} \overline{\mathcal{B}_{z_k,\varepsilon}}) = J_\varepsilon(u_\varepsilon),
\tag{19}
$$

Our identification problem can be formulated as a topological optimization problem one. It consists in finding the optimal location of the holes $\mathcal{B}_{z_k,\varepsilon} = z_k + \varepsilon \mathcal{B}^k$, $1 \leq k \leq m$, inside the domain Ω in order to minimize the optimal design function j.

$$
(\mathcal{O}_\varepsilon) \begin{cases}
\text{Find } z_k^* \in \Omega, 1 \leq k \leq m, \text{ such that :} \\
j(\Omega \backslash \cup_{k=1}^{m} \overline{\mathcal{B}_{z_k^*,\varepsilon}}) = \min_{\mathcal{B}_{z_k,\varepsilon} \subset \Omega} j(\Omega \backslash \cup_{k=1}^{m} \overline{\mathcal{B}_{z_k,\varepsilon}}).
\end{cases}
$$

To solve this optimization problem $(\mathcal{O}_\varepsilon)$ we have used the topological sensitivity analysis method. It consists in studying the variation of the design function j with respect to the presence of a small injector $\mathcal{B}_{z,\varepsilon} = z + \varepsilon \mathcal{B}$ in the lake Ω.

4.3.2 Numerical results

We propose an adaptation of the previous algorithm to our context. At the kth iteration, the topological gradient δj_k is given by

$$
\delta j_k(z) = \left(u_k(x) - u_{inj} \right) \cdot v_k(x), \quad \forall z \in \Omega_k
\tag{20}
$$

where u_k and v_k are, respectively, solutions to the direct and adjoint problems in Ω_k.
We consider the set $\{x \in \Omega_k; \quad \delta j_k(x) < c_{k+1}\}$. Each connected component of this set is a hole created by the algorithm. Our idea is to replace each hole by an injector located at the local minimum of $\delta j_k(x)$. The obtained results are described in figures 16 and 17.

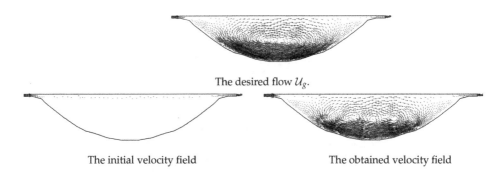

The desired flow $\mathcal{U}_g$.

The initial velocity field   The obtained velocity field

Fig. 16. Numerical results in 2D (for more details one can see Hassine and Masmoudi (2004)).

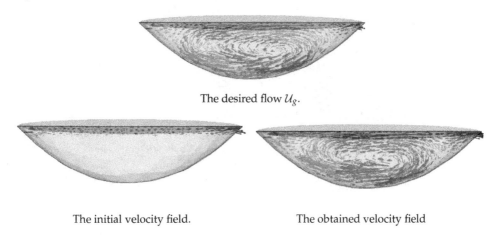

The desired flow $\mathcal{U}_g$.

The initial velocity field.   The obtained velocity field

Fig. 17. Numerical results in 3D (see Abdelwahed, Hassine and Masmoudi (2009)).

4.4 Geometrical control of fluid flow

We consider a tank Ω filled with a viscous and incompressible fluid. The aim is to determine the optimal shape of the fluid flow domain minimizing a given objective function.

4.4.1 Approximation of a desired flow

The aim is to determine the optimal shape $\mathcal{O}^* \subset \Omega$ of the fluid flow domain such that the velocity $u_{\mathcal{O}^*}$, solution to the Stokes equations in $\mathcal{O}^*$, approximate a desired flow w_d defined in a fixed domain Ω_m. The optimal shape $\mathcal{O}^*$ can be characterized as the solution to the following topological optimization problem

$$\min_{D \subset \Omega} \int_{\Omega_m} |u_D - w_d|^2 dx,$$

where u_D is the solution to the Stokes equations in $D \subset \Omega$. This test is treated in two and three dimensional cases. In 2D, the tank $\Omega = [0, 1.5] \times [0, 1]$, the domain $\Omega_m = [0, 1.5] \times [0.8, 1]$ and the velocity field w_d is defined by

$$w_d = \begin{cases} (1, 0) \text{ in } \Omega_m, \\ (0, 0) \text{ elsewhere .} \end{cases}$$

The numerical results are described in Figure 18. A 3D extension of this case is presented in Figure 19.

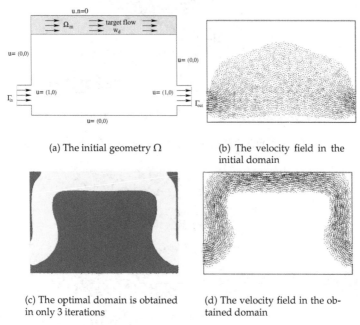

(a) The initial geometry Ω

(b) The velocity field in the initial domain

(c) The optimal domain is obtained in only 3 iterations

(d) The velocity field in the obtained domain

Fig. 18. Approximation of a desired flow: 2D case

4.4.2 Maximizing velocity in a fixed zone

Here the aim is to maximize the fluid flow velocity in $\Omega_m = \cup_k \Omega_m^k \subset \Omega$ (fixed zones) using a topological perturbation of the domain. The optimal domain of the fluid flow can be characterized as a solution to the following problem

$$\max_{\mathcal{O} \subset \Omega} \int_{\Omega_m} |u_{\mathcal{O}}|^2 dx,$$

where $u_{\mathcal{O}}$ is the solution to the Stokes equations in $\mathcal{O}$.

Two 3D test cases are considered. The first case is described in Figure 20. The inflow Γ_{in} and the outflow Γ_{out} (see Figure 20(a)) are defined by

$\Gamma_{in} = [0, 1.5] \times 0 \times [0.4, 0.6]$, $\Gamma_{out} = [0, 1.5] \times 0 \times [0.4, 0.6]$.

The domain $\Omega_m = \Omega_m^1 \cup \Omega_m^2$, with $\Omega_m^1 = [0, 1.5] \times [0, 1] \times [0.9, 1]$ and $\Omega_m^2 = [0, 1.5] \times [0, 1] \times [0, 0.1]$.

The optimal domain (see Figure 20(c)) is obtained in four iterations.

The second case is described in Figure 21. Here we have used the same 3D tank considered in the last case but with different Γ_{in}, Γ_{out} and Ω_m (see Figure 21(a)). The optimal domain (see Figure 21(c)) is obtained in five iterations.

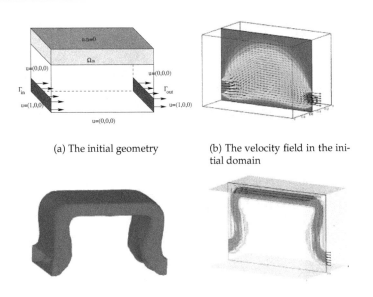

(a) The initial geometry

(b) The velocity field in the initial domain

(c) The optimal domain is obtained in only 4 iterations

(d) The velocity field in the obtained domain

Fig. 19. Approximation of a desired flow: 3D case

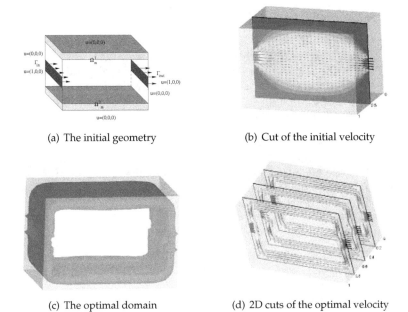

(a) The initial geometry

(b) Cut of the initial velocity

(c) The optimal domain

(d) 2D cuts of the optimal velocity

Fig. 20. Maximizing velocity in a fixed zone: first case (see Abdelwahed and Hassine (2009))

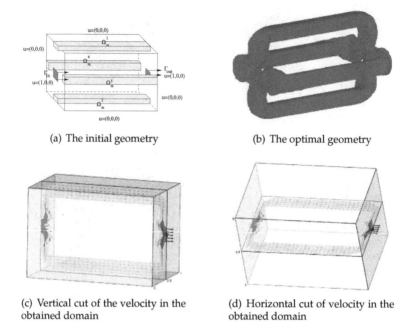

(a) The initial geometry

(b) The optimal geometry

(c) Vertical cut of the velocity in the
obtained domain

(d) Horizontal cut of velocity in the
obtained domain

Fig. 21. Maximizing velocity in a fixed zone: second case (see Abdelwahed and Hassine (2009))

5. Conclusion

In this chapter we have proposed an accurate and fast topological optimization algorithm. The optimal domain is obtained iteratively by inserting some obstacles at each iteration. The location and size of the obstacles are described by a scalar function called the topological gradient. The topological gradient is derived as the leading term of the cost function variation with respect to the insertion of a small obstacle in the fluid flow domain.

The proposed method has two main features. The first one concerns its mathematical framework. The topological sensitivity analysis can be adapted for various operators like elasticity, Helmholtz, Maxwell, Navier Stokes, ...

The second interesting feature of the approach is that it leads to a fast and accurate numerical algorithm. Only a few iterations are needed to construct the final domain. It is easy to be implemented and can be used for many applications. At each iteration we only need to solve the direct and the adjoint problems on a fixed grid.

6. References

M. Abdelwahed and M. Hassine, *Topological optimization method for a geometric control problem in Stokes flow*, Journal of Applied Numerical Mathematics, Volume 59 (8), 2009, 1823-1838.

M. Abdelwahed, M. Hassine and M. Masmoudi, *Optimal shape design for fluid flow using topological perturbation technique*, Journal of Mathematical Analysis and Applications, Volume 356, 2009, 548–563.

M. Abdelwahed, M. Hassine and M. Masmoudi, *Control of a mechanical aeration process via topological sensitivity analysis,* Journal of Computational and Applied Mathematics, Volume 228 (1), 2009, 480–485.

G. Allaire and R. Kohn, *Optimal bounds on the effective behavior of a mixture of two well-orded elastic materials,* Quartely of Applied Mathematics, Li(4), 1996, 643-674.

A. Ben Abda, M. Hassine, M. Jaoua, M. Masmoudi, *Topological sensitivity analysis for the location of small cavities in Stokes flow,* SIAM J. Contr. Optim. Vol. 48 (5), 2009, 2871–2900

S. Amstutz, *The topological asymptotic for Navier Stokes equations,* ESAIM, Cont. Optim. Cal. Var. 11 (3), 2005, 401-425.

S. Amstutz, M. Masmoudi and B. Samet, *The topological asymptotic for the Helmholtz equation,* SIAM J. Contr. Optim, 42 (2003), 1523-1544.

D. Arnold, F. Brezzi and M. Fortin (1984), *A stable finite element for the Stokes equations,* Calcolo, 21 (4) (1984), 337-344.

M. Bendsoe, *Optimal topology design of continuum structure: an introduction.* Technical report, Departement of mathematics, Technical University of Denmark, DK2800 Lyngby, Denmark, september 1996.

T. Borrvall and J. Petersson, *Topological optimization of fluids in Stokes flow,* Inter. J. Numer. Methods Fluids, 41 (1), 2003, 77-107.

G. Buttazzo and G. Dal Maso, *Shape optimization for Dirichlet problems: Relaxed formulation and optimality conditions,* Appl. Math. Optim. 23 (1991), 17-49.

H. Cabuk and V. Modi, *Optimum plane diffusers in laminar flow,* J. of Fluid Mechanics 237 (1992), 373-393.

J. Céa, S. Garreau, Ph. Guillaume and M. Masmoudi, *The shape and Topological Optimizations Connection,* Comput. Methods Appl. Mech. Engrg. 188(4), (2000) 713-726.

E.M. Cliff, M. Heinkenschloss and A. Shenoy, *Airfoil design by an all-at-once method,* Inter. J. Compu. Fluid Mechanics 11 (1998) 3-25.

S.S. Collis, K. Ghayour, M. Heinkenschloss, M. Ulbrich and S. Ulbrich, *Optimal control for unsteady compressible viscous flows,* Inter. J. Numer. Meth. Fluids 40 (2002) 1401-1429.

R. Dautray and J. Lions, *Analyse mathémathique et calcul numérique pour les sciences et les techniques,* MASSON, collection CEA, 1987.

A. Evgrafov, *The Limits of Porous Materials in the Topology Optimization of Stokes Flows,* Applied Mathematics and Optimization 52 (3), (2005) 263-277.

S. Garreau, Ph. Guillaume and M. Masmoudi, *The topological asymptotic for pde systems: the elasticity case,* SIAM J. Control Optim., 39(4) (2001) 1756-1778.

D.K. Gartling, C.E.Hickox and R.C. Givler, *Simulation of coupled viscous and porous flow problems,* Comp. Fluid. Dyn., 7(1) (1996) 23-48.

O. Ghattas and J-H. Bark, *Optimal control of two and three dimensional incompressible Navier-Stokes flows,* Journal of Computational Physics, 136 (1997) 231-244.

R. Glowinski and O. Pironneau, *On the numerical computation of the minimum drag profile in laminar flow,* J. of Fluid Mechanics 72 (1975) 385-389.

J. K. Guest and J. H. Prévost, *Topology optimization of creeping fluid flows using a Darcy-Stokes finite element,* Inter. J. Numer. Methods in Engineering 66 (2006) 461-484.

Ph. Guillaume and M. Hassine, *Removing holes in topological shape optimization*, ESAIM, COCV J. vol. 14 (1) (2007) 160-191.

Ph. Guillaume and K. Sid Idris, *Topological sensitivity and shape optimization for the Stokes equations.* SIAM J. Control Optim. 43(1) (2004) 1-31.

M.D. Gunzburger *Perspectives in flow control and optimization*, Advances in Design and Control, SIAM, Philadelphia, PA, 2003.

M.D. Gunzburger, L. Hou and T. Sovobodny, *Analysis and finite element approximation of optimal control problems for the stationary Navier-Stokes equations with Dirichlet controls*, RAIRO Model. Math. Anal. Numer. 25 (1991) 711-748.

M. Gunzburger, H. Kim and S. Manservisi, *On a shape control problem for the stationary Navier-Stokes equations*, M2AN Math. Model. Numer. Anal. 34 (6) (2000) 1233-1258.

M. Hassine, *Shape optimization for the Stokes equations using topological sensitivity analysis*, ARIMA Journal, vol. 5, (2006) 213-226.

M. Hassine, *Topological Sensitivity Analysis: theory and applications*, Habilitation Universitaire, El Manar University, Tunisia, 2008.

M. Hassine, S. Jan and M. Masmoudi, *From differential calculus to* $0-1$ *topological optimization*, SIAM J. Cont. Optim. vol.45 (6) (2007) 1965-1987.

M. Hassine and M. Masmoudi, *The topological asymptotic expansion for the Quasi-Stokes problem*, ESAIM, COCV J. vol. 10 (4) (2004) 478-504.

M. Ishii, *Thermo-fluid dynamic theory of a two-phase flow*, Collection de la direction des études de recherche d'électricité de france, EYROLLES, 1975.

D.W. Kim and M.U. Kim, *Minimum drag shape in two two-dimensional viscous flow*, Inter. J. Numer. Methods in Fluids 21 (1995) 93-111.

E. Lund, H. Moller and L.A. Jakobsen, *Shape Optimization of Fluid-Structure Interaction Problems Using Two-Equation Turbulence Models.* In Proc. 43rd Structures, Structural Dynamics, and Materials Conference and Exhibit, 2002, Denver, Colorado (CDROM), AIAA 2002-1478, 11 pages.

M. Masmoudi, *The topological asymptotic*, In Computational Methods for Control Applications, ed. H. Kawarada and J. Periaux, International Séries GAKUTO, 2002.

B. Mohammadi and O. Pironneau, *Applied shape optimization for fluids*, Numerical Mathematics and Scientific Computation, Oxford University Press, New York, 2001.

O. Pironneau, *On the transport-diffusion algorithm and its applications to the Navier-Stokes equations*, Numerische Mathematik, 38 (1982), 309-332.

O. Pironneau, *On optimum profiles in Stokes flow*, J. of Fluid Mechanics 59 (1973) 117-128.

O. Pironneau, *On optimum design in fluid mechanics*, J. of Fluid Mechanics 64 (1974) 97-110.

O. Pironneau, *Optimal Shape Design for Elliptic Systems*, Springer, Berlin (1984).

J. Simon, *Domain variation for drag Stokes flows*. In Lecture Notes in Control and Inform. Sci. 114, A. Bermudez Eds. Springer, Berlin (1987) 277-283.

A. Schumacher, *Topologieoptimierung von bauteilstrukturen unter verwendung von lopchpositionierungkrieterien*, thesis, Universitat-Gesamthochschule-Siegen, 1995.

J. Sokolowski and A. Zochowski, *On the topological derivative in shape optimization*, SIAM J. Control Optim., 37 (4) (1999) 1251-1272.

10

Multiscale Particle-In-Cell Method: From Fluid to Solid Mechanics

Alireza Asgari[1] and Louis Moresi[2]

[1]*School of Engineering, Deakin University,*
[2]*School of Mathematical Sciences, Monash University,*
Australia

1. Introduction

In this chapter, a novel multiscale method is presented that is based upon the Particle-In-Cell (PIC) finite element approach. The particle method, which is primarily used for fluid mechanics applications, is generalized and combined with a homogenization technique. The resulting technique can take us from fluid mechanics simulations seamlessly into solid mechanics. Here, large deformations are dealt with seamlessly, and material deformation is easily followed by the Lagrangian particles in an Eulerian grid. Using this method, solid materials can be modeled at both micro and macro scales where large strains and large displacements are expected. In a multiscale simulation, the continuum material points at the macroscopic scale are history and scale dependent. When conventional Finite Elements are used for multiscale modeling, microscale models are assigned to each integration point; with these integration points usually placed at Gaussian positions. Macroscale stresses are extrapolated from the solution of the microscale model at these points and significant loss of information occurs whenever re-meshing is performed. In addition, in the case of higher-order formulations, the choice of element type becomes critical as it can dictate the performance, efficiency and stability of the multiscale modeling scheme. Higher-order elements that provide better accuracy lead to systems of equations that are significantly larger than the system of equations from linear elements. The PIC method, on the other hand, avoids all of the problems that arise from element layout and topology because every material point carries material history information regardless of the mesh connectivity. Material points are not restricted to Gaussian positions and can be dispersed in the domain randomly, or with a controlled population and dispersion. Therefore, in analogy to Finite Element Method (FEM) mesh refinement, parts of the domain that might experience localization can have a higher number of material points representing the continuum macroscale in more detail. Similar to the FEM multiscale approach, each material point has a microscale model assigned and information passing between the macro and micro scales is carried out in a conventional form using homogenization formulations. In essence, the homogenization formulation is based on finding the solution of two boundary value problems at the micro and macro scales with information passing between scales. Another advantage of using the particle method is that each particle can represent an individual material property, thus in the microscale phase, interfaces can be followed without the numerical overhead associated with the contact algorithms while material points move. Moreover, the PIC method in the macro and microscale tracks material discontinuities and can model highly distorted material flows. Here we aim to build and investigate the applicability of the PIC method in a multiscale framework.

Three examples that showcase a fluid approach to model solid deformation are presented. The first numerical experiment uses a viscoelastic-plastic material that undergoes high strains in shear. The material is considered to be in a periodic domain. The initial orientation, the high-strain configuration and the softening of the shear bands are studied with very high resolutions using PIC method. This is a situation where a conventional FE approach could be very expensive and problematic. The second example is the simulation of the global system of tectonic plates that form the cool upper boundary layer of deep planetary convection cells. Creeping flow in the mantle drives the plates and helps to cool the Earth. This is an unusual mode of convection, not only because it takes place in a solid silicate shell, but also because a significant fraction of the thickness of the cold boundary layer (the lithosphere) behaves elastically with a relaxation time comparable to the overturn time of the system (Watts et al. 1980). In modeling this system, we need to account for Maxwell viscoelastic behavior in the lithosphere as it is recycled into the Earth's interior. The lithosphere undergoes bending, unbending, and buckling during this process, all the while being heated and subjected to stress by the mantle flow. A Maxwell material has a constitutive relationship which depends upon both the stress and the stress rate. The stress rate involves the history of the stress tensor along particle trajectories. PIC approaches are an ideal way to approach this problem. For a full discussion, see (Moresi et al. 2003). The last example presented in this chapter is on the simulation of forming process of Advanced High Strength Steels (AHSS). These are materials with multiphase microstructure in which microstructure has a pronounced effect on the forming and spring-back of the material. The contribution of this chapter is on the illustration of the PIC method in solid mechanics through examples that include large strain deformation of viscoelastic-plastic materials as well as multiscale deformation of elastoplastic materials.

2. Particle-based method: An alternate to FE method

Particle methods are result of attempts to partially or completely remove the need for a mesh. In these methods, an approximation to the solution is constructed strictly in terms of nodal point unknowns (Belytschko et al. 1996). The domain of interest is discretized by a set of nodes or particles. Similar to FEM, a shape function is defined for each node. The region in the function's support, usually a disc or rectangle, is called the domain of influence of the node (Belytschko et al. 1996). The shape function typically has two parameters, providing the ability to translate and dilate the domain of influence of a shape function. The translation parameter allows the function to move around the domain, replacing the elements in a meshed method. The dilation parameter changes the size of the domain of influence of the shape function, controlling the number of calculations necessary to find a solution. As the dilation parameter becomes larger, larger time steps can be taken. A set of basis functions also needs to be defined for a given problem.

The principle advantage of a particle method is that the particles are not treated as a mesh with a prescribed connectivity. Therefore, mesh entanglement is not a problem and large deformations can be treated easily with these methods. Creating new meshes and mapping between meshes is eliminated. Refinement can be obtained by simply adding points in the region of interest (Liu et al. 2004). It should be noted that most of these methods use a mesh to perform integration. However, this mesh can be simpler than would be needed for standard element-based solutions. Another advantage of the particle method is that there is no need to track material interfaces, since each particle has its own constitutive properties (Benson 1992).

There are also difficulties in using particle methods. One of the major disadvantages of these methods is their relatively high computational cost, particularly in the formation of the stiffness matrix. The support of a shape function generally overlaps surrounding points unlike standard FE shape functions. In fact, the support of the kernel function must cover a minimum number particles for the method to be stable (Chen et al. 1998).

The bandwidth of the stiffness matrix is increased by the wider interaction between nodes. There is a more irregular pattern of the sparsity, since the number of neighbors of a given point can vary from point to point. Thus, the number of numerical operations in the formation and application of these matrices increases. Additionally, higher-order shape and basis functions are usually used, with the result that higher-order integrations are required. Construction of these shape functions is also costly (Chen et al. 1998).

Another problem is that overlapping shape functions are not interpolants (Belytschko et al. 2001); even the value of a function at a nodal point must be computed from the contributing neighbors. This requirement makes essential boundary conditions more difficult to apply. Some techniques that have been developed to address this problem are Lagrange multipliers, modified variational principles, penalty methods and coupling to FEMs (Belytschko et al. 2001). However, there can also be difficulties with using these techniques. For example, the Lagrange multiplier method requires solution of an even larger system of equations. In addition, Lagrangian multipliers tend to destroy any structure, such as being banded or positive definite, that the system might exhibit (Belytschko et al. 2001). The modified variational approach applies boundary conditions of a lower order of accuracy. Coupling to FEMs by using particle methods only in regions with large deformations and FEMs elsewhere in the problem can reduce the cost of the solution. However, the shape functions at the interface become quite complicated and require a higher order of quadrature (Belytschko et al. 2001). Another method (Chen et al. 1998) first uses the map from nodal values to the function space to get nodal values of the function, then applies the boundary conditions, and then transforms back to nodal values.

Many of these codes have been restricted to static problems. Chen has developed a dynamic code, but all of the basis functions are constructed in the original configuration (Chen et al. 1998).

Like their meshed counterparts, particle methods are useful for certain types of problems. However, because of their higher computational cost, they are not the method of choice for some types of problems. The PIC method, also known as Material Point Method (MPM), combines some aspects of both meshed and meshless methods. This method was developed by Sulsky and co-workers for solid mechanics problems based on the Fluid-Impact-Particle (FLIP) code, which itself was a computational fluid dynamics code (Sulsky et al. 1995).

Sulsky et al. (Sulsky et al. 1995) initially considered application of MPM to 2D impact problems and demonstrated the potential of MPM. Later in 1996, Sulsky and Schreyer (Sulsky and Schreyer 1996) gave a more general description of MPM, along with special considerations relevant to axisymmetric problems. The method utilizes a material or Lagrangian mesh defined on the body under investigation, and a spatial or Eulerian mesh defined over the computational domain. The advantage here is that the set of material points making up the material mesh is tracked throughout the deformation history of the body, and these points carry with them a representation of the solution in a Lagrangian frame. Interactions among these material points are computed by projecting information they carry onto a background FE mesh where equations of motion are solved. Furthermore, the MPM does not exhibit locking or an overly stiff response in simulations of upsetting (Sulsky and Schreyer 1996).

In MPM, a solid body is discretized into a collection of material points, which together represent the body. As the dynamic analysis proceeds, the solution is tracked on the material points by updating all required properties, such as position, velocity, acceleration, stress, and temperature. At each time step, the material point information is extrapolated to a background grid, which serves as a computational scratch pad to solve the equilibrium equations. The solution on the grid-nodes is transferred using FE shape functions, on the material points, to update their values. This combination of Lagrangian and Eulerian methods has proven useful for solving solid mechanics problems involving materials with history dependent properties, such as plasticity or viscoelastic effects.

The key difference between The PIC method developed by Moresi et al. and the classical MPM is the computation of updated particle weights which differ from the initial particle mass. This gives much improved accuracy in the fluid-deformation limits especially for incompressible flows (Moresi et al. 2003). This method is amendable to efficient parallel computation and can handle large deformation for viscoelastic materials. In the next section, we explain how this method is combined with the localization and homogenization routines to build a multiscale PIC method suitable solid mechanics applications.

3. Numerical formulation

3.1 Incremental iterative algorithm

The stiffness matrix term is assembled over all cells ($K = A_{e=1}^{ncell} k$) in the model where the stiffness matrix of each cell is

$$k = \int_{\Omega^e} B^T \, CB \, d\Omega \tag{1}$$

In equation (1) C is a fourth order tensor that must be calculated by the appropriate constitutive equations at the particle level. In calculation of the global stiffness matrix, the effect of initial stresses (or nonlinear geometrical effects) must be considered. This is achieved by selecting a proper reference frame and adding nonlinear strain-displacement terms to the calculation of equation (1).

We choose to use an Updated Lagrangian reference frame to build the global stiffness matrix. For simplicity, a quasi-static condition is considered. Therefore, the linearized system of equations takes the following form

$$(K_{\text{Linear}} + K_{\text{Nonlin}})U = F_{\text{ext}} - F_{\text{int}} \tag{2}$$

Similar to FEM matrix and vector assembly, Equation (2) in extended form can be written as

$$\left(\int_v B_{\text{Linear}}^T \, CB_{\text{Linear}} \, dv + \int_v B_{\text{Nonlin}}^T \, \tau B_{\text{Nonlin}} \, dv \right) u = F_{\text{ext}}^{t+\Delta t} - \int_v B_{\text{Linear}}^T \, \tau \, dv \tag{3}$$

In equation (3), B_{Linear} and B_{Nonlin} are linear and nonlinear strain-displacement matrices and the external forces are contributions of body forces and tractions

$$F_{ext}^{t+\Delta t} = \int_s N^T \, fs_0^{t+\Delta t} ds + \int_v N^T \, fb_0^{t+\Delta t} dv \tag{4}$$

In Equations (2)-(4), terms are referenced to time t and $t+\Delta t$, whereas if the Total Lagrangian frame were in use, terms with reference to time zero and t should have been taken into account. Similarly, the global internal force vector is assembled over all cells ($\Delta F_{int} = A_{e=1}^{ncell} \Delta f_{int}$). The difference between the internal and external force vectors is the residual, or out-of-balance, term:

$$R = \Delta F_{int} - \Delta F_{ext} \tag{5}$$

If $R < \lambda_{tolerance}$ then the current increment $(n + 1)^{th}$ is in equilibrium and the simulation can move forward to the next increment, otherwise the system of linear equation is modified such that iterative displacements can be found from

$$KdU = R. \tag{6}$$

With each new iterative displacement solution dU, the above procedure is repeated until the criteria of $R < \lambda_{tolerance}$ is satisfied (convergence is achieved). The preceding algorithm is the standard incremental-iterative algorithm, which can be used for both macro and micro scale simulation. The important factor to consider in the multiscale PIC is the coupling between these two scales and the constitutive relations at each scale.

Constitutive equations of different phases at the microscale are defined using a conventional continuum mechanics approach for elastoplastic materials.

3.2 Integration using particles
Following the works of Moresi et al. (2003), the sampling points for integration of field variables in our multiscale PIC method are matched with the material points embedded in the macro scale model. These integration points are not fixed like Gauss points. Therefore, their positions are not known in advance and an adaptive scheme with the procedure outlined in the numerical implementation of Moresi et al. (2003) is used. The integral in Equation (1) results in a system of linear equations with unknown displacements on nodes with the element stiffness matrix

$$k_{ij}^e = \sum_{p=1}^{n_{ep}} w_p B_i^T(x_p) C_p B_j^T(x_p) \tag{7}$$

where n_{ep} is the number of particles with weight w_p in a cell. B is the usual linear strain-displacement tensor in FE. The fourth order tensor C_p is the stiffness matrix of the material at any arbitrary point P. The core significance of the multiscale PIC method lies in the way C_p is calculated and obtained. In single scale methods, the constitutive equations are used to build this matrix. However, we use microscale homogenization, also known as multi level FE, to build this tensor as described shortly. An incremental iterative solution strategy of full Newton-Raphson (N-R) is used to solve the linearized equations of motion for nodal data in an implicit manner (Guilkey and Weiss 2003). The effect of initial stress and nonlinear geometric effect (e.g. due to mismatch between material properties of particles) is taken into account by addition of a nonlinear term to the stiffness matrix in equation (7). In all of the analyses, a quasi-static situation is assumed and all body and surface forces are represented by the external loads. After finding the incremental displacement, the strain and deformation gradient for each particle are calculated and used for the macro-micro coupling.

3.3 Elastoplastic solid deformation

In continuum and fluid mechanics, the underlying microscopic mechanisms are complex but simplified at the macroscale level. In multiscale PIC, classical elastoplasticity is used as the constitutive relations for the microscale phases.

Transformation from the undeformed configuration at pseudo time t_0 to the current configuration at pseudo time t is described by a deformation gradient tensor F. The total deformation is multiplicatively decomposed into elastic and plastic parts. The decomposition is unique with the common assumption that the plastic rotation rate during the current increment is zero. Therefore, material rotations are fully represented in the elastic part of the deformation gradient tensor.

The von Mises yield criterion defined by

$$f(\sigma, \bar{\varepsilon}_p) = \frac{3}{2}\sigma' : \sigma' - \sigma_y^2(\bar{\varepsilon}_p) \tag{8}$$

is used, where the effective plastic strain is

$$\bar{\varepsilon}_p = \int_{\tau=0}^{t} \sqrt{\frac{2}{3}D_p : D_p}\, d\tau \tag{9}$$

Hydrostatic stresses σ' are calculated using mean stress

$$\sigma_m = \frac{\sigma_{ii}}{3}$$

$$\sigma'_{ij} = \sigma_{ij} - \sigma_m \delta_{ij} \tag{10}$$

During the elastoplastic deformation, the plastic deformation rate D_p is related to the stress by the normality rule. The direction of D_p is perpendicular to the yield surface in the stress space, and the length of D_p, which is unknown at the current increment, is characterized by the plastic multiplier $\dot{\lambda}$. The value of the plastic multiplier is found from the consistency equation so that the stress state always resides on the yield surface during elastoplastic deformation. When the plastic multiplier is calculated, the constitutive relation between the objective rate of the Cauchy stress tensor and the deformation rate tensor is determined using the Prandtl-Reuss equation. Besides the constitutive equation for the stress state, the effective plastic strain is updated to keep track of the plastic evolution.

3.4 Viscoelastic fluid deformation

For the viscoelastic deformation, we begin our analysis from the momentum conservation equation of an incompressible, Maxwell viscoelastic fluid having infinite Prandtl number (i.e. all inertial terms can be neglected), and driven by buoyancy forces.

$$\sigma_{ij,j} = f_i \tag{11}$$

where σ is the stress tensor and f a force term. The stress consists of a deviatoric part, τ, and an isotropic pressure, p,

$$\sigma_{ij} = \tau_{ij} - p\delta_{ij} \tag{12}$$

where $\delta_{ij} = I$ is the identity tensor.

The Maxwell model assumes that the strain rate tensor, D, defined as:

$$D_{ij} = \frac{1}{2}\left(\frac{\partial v_i}{\partial x_j} + \frac{\partial v_j}{\partial x_i}\right) \tag{13}$$

is the sum of an elastic strain rate tensor D^e and a viscous strain rate tensor D^v. The velocity vector, V, is the fundamental unknown of our problem and all these entities are expressed in the fixed reference frame x_i. In an incompressible material, these strain rates are, by definition, devatoric tensors.

The viscous and elastic constitutive laws in terms of the strain rate are, respectively

$$\tau_{ij} = 2\eta D^v; \quad \check{\tau}_{ij} = 2\mu D^e \tag{14}$$

where η is shear viscosity, μ is the shear modulus, and $\check{\tau}$ is an objective material derivative of the deviatoric stress.

The viscous and elastic constitutive laws are combined by summing each contribution to the strain rate tensor

$$D_{ij} = D^e_{ij} + D^v_{ij} = \frac{\check{\tau}_{ij}}{2\mu} + \frac{\tau_{ij}}{2\eta} \tag{15}$$

Writing the stress derivative as the difference, in the limit of small $t - \Delta t_e$, between the current stress solution τ at t, and the stress at an earlier time, $t - \Delta t_e$ gives

$$\check{\tau}_{ij}(t, x) = \lim_{\delta \Delta t_e \to 0} \frac{\tau^t_{ij}(t,x) - \hat{\tau}^t_{ij}(t, t-\Delta t_e, x, u(x,t))}{\delta t} \tag{16}$$

Where $\hat{\tau}^t_{ij}$ indicates a stress history tensor which has been transported to the current location by the time-dependent velocity field from its position at the earlier time, $t - \Delta t_e$. For each point in the domain, x, this stress history is dependent upon the velocity history, which determines the path taken by the material to reach the point, x, and any rotation of the tensor along the path.

A common choice for $\check{\tau}$ is the following:

$$\check{\tau}_{ij} = \frac{\partial \tau_{ij}}{\partial t} + \tau'_{ij} \tag{17}$$

where τ'_{ij} is the instantaneous rate of change in the stress tensor, related to the spatial part of the Eulerian derivative and associated with the transport, rotation and stretching by fluid motion.

$$\tau'_{ij} = u_k\frac{\partial \tau_{ij}}{\partial x_k} + \tau_{ik}W_{kj} - W_{ik}\tau_{kj} + a(\tau_{ik}D_{kj} + D_{ik}\tau_{kj}) \tag{18}$$

Here a is a parameter that can take values between -1 and 1, and W is the spin tensor,

$$W_{ij} = \frac{1}{2}\left[\frac{\partial v_i}{\partial x_j} - \frac{\partial v_j}{\partial x_i}\right] \tag{19}$$

When $a = 0$, $\check{\tau}$ is known as the Jaumann derivative, and when $a = 1$ or $a = -1$ it is known as the upper- or lower-convected Maxwell derivative, respectively. In our implementation, we choose $a = 0$. This choice greatly simplifies the formulation as it ensures that $\check{\tau}$ is deviatoric when τ is deviatoric.

Numerically, with the PIC formulation, it is more convenient to compute $\hat{\tau}$ directly than to attempt to compute the derivatives in τ'. $\hat{\tau}$ is the stress tensor stored at a material point accounting for rotation along the path. This gives the following form of the constitutive relationship.

$$\tau_{ij} = 2\frac{\eta\mu\Delta t_e}{\eta+\mu\Delta t_e}\left[D_{ij} + \frac{\hat{\tau}_{ij}^{\Delta t_e}}{2\mu\Delta t_e}\right] \tag{20}$$

If we now write

$$\eta^{\text{eff}} = \frac{\eta\mu\Delta t_e}{\eta+\mu\Delta t_e} \tag{21}$$

$$D_{ij}^{\text{eff}} = \left[D_{ij} + \frac{\hat{\tau}_{ij}^{\Delta t_e}}{2\mu\Delta t_e} -\right] \tag{22}$$

$$\tau_{ij} = 2\eta^{\text{eff}}D_{ij}^{\text{eff}} \tag{23}$$

It is clear that the introduction of viscoelastic terms into a fluid dynamics code is merely a question of being able to compute $\hat{\tau}$ accurately; a simple task when particles are available.

An important feature of this formulation is the fact that we have specified an independent *elastic time step,* Δt_e which is not explicitly related to the relaxation time of the material or the timestep mandated by the Courant stability criterion.

The use of an independent step is appropriate because there is a large variation of relaxation time within the lithosphere due to temperature dependence of viscosity exhibited by mantle rocks (Karato 1993). While the near surface has a relaxation time comparable to the overturn time, the deeper, warmer parts of the lithosphere have a relaxation time several orders of magnitude less than this — and elasticity may be ignored altogether. If Δt_e is chosen to resolve elastic effects at the relaxation time of the cool part of the boundary layer, then, for regions where the relaxation time is much smaller, the elastic term simply becomes negligibly small and viscous behavior is recovered. In (Moresi et al. 2003), we describe the manner in which the stress tensor can be averaged over a moving window to ensure that the storage of only one history value is required to recover $\hat{\tau}$ at $t - \Delta t_e$.

4. Multiscale formulation

To perform multiscale simulation, a Representative Volume Element (RVE) is assigned to each material point. Instead of constitutive relations at the macroscale level, the microstructural material properties and morphologies are assumed to be known *a priori*. The realization of the RVE in terms of size and representativeness is an important factor. At the RVE level, the macroscale deformation gradient is used to impose boundary conditions on the RVE.

After solving the microscale problem using a stand-alone multiscale PIC model, the stresses and tangent stiffness matrix are averaged and returned to the macroscale material point. The localization and homogenization (or commonly known the multilevel FEM) approaches explained here are mainly inspired by the works of Smit et al. (1998) and Kouznetsova et al. (2001).

After performing the homogenization, the residual forces at the material points are calculated. Then convergence of the macroscale iteration is checked. If convergence is achieved, the solver proceeds to the next increment; otherwise, the microscale solution is initiated to obtain the corrective stress vector. The material tangent-stiffness matrix and the averaged stress vector are evaluated for the corresponding material point.

4.1 Localization
During this phase, Boundary Conditions (BCs) are applied on the RVE model. BCs can be imposed in three ways: a) by prescribed displacement, b) by prescribed traction, or c) by periodic boundary conditions. The following formulation covers the prescribed displacement and periodic BC cases. The formulation for traction BC is outside the scope of our multiscale PIC code because it is fundamentally a displacement driven code.

The macroscale deformation gradient is used as a mapping tensor on boundaries of the RVE, $x = F_{\text{Macro}} \cdot X$ where x refers to deformed and X to undeformed position vectors of the points along the boundary. The most reasonable estimation of the averaged properties are obtained with periodic BCs (Smit et al. 1998). The periodic BCs are enforced using symmetric displacement and antisymmetric traction BCs along opposite edges of the boundary $\Gamma_{ij\,0}$ (Kouznetsova et al. 2001).

4.2 Homogenization
The coupling of the macro and micro scales is performed during the homogenization phase. In this phase of solution, the averaged stress and consistent tangent stiffness matrix are calculated for the multiscale constitutive relations using the computational homogenization method outline by Kouznetsova (Kouznetsova 2002; Kouznetsova et al. 2004).

The macroscale deformation gradient and 1st PK stress are volume averages of their microscale counterparts.

$$F_{\text{Macro}} = \frac{1}{v} \int_{V_0} F_{\text{Micro}} dV_0 \tag{24}$$

$$P_{\text{Macro}} = \frac{1}{v} \int_{V_0} P_{\text{Micro}} dV_0 \tag{26}$$

where P is the traction or force. After the RVE microscale problem is solved, the values of P_{Micro} are known, and one can perform the integration in equation (26) and return macroscale stresses to the corresponding material point. In the following, the gradient operator ∇_0 is taken with respect to reference configuration while ∇ is the gradient operator with respect to the current configuration. With account for microscale equilibrium $\nabla_{0\,\text{Micro}} \cdot P_{\text{Micro}} = \vec{0}$ and the equality of $\nabla_{0\,\text{Micro}} \cdot X = \vec{I}$ the following relation holds

$$P_{\text{Micro}} = (\nabla_{0\,\text{Micro}} \cdot P_{\text{Micro}})X + P_{\text{Micro}}(\nabla_{0\,\text{Micro}} \cdot X) = \nabla_{0\,\text{Micro}}(P_{\text{Micro}} \cdot X) \tag{27}$$

substitution of equation (27) into (26) and application of the divergence theorem ($\int_\Omega \nabla \cdot u d\Omega = \int_\Gamma n \cdot u\, d\Gamma$) and the definition of the first PK stress vector ($p = N \cdot P_{\text{Micro}}$) gives

$$P_{\text{Macro}} = \frac{1}{v} \int_{V_0} \nabla_{0\,\text{Micro}} \cdot (P_{\text{Micro}}X) dV_0 = \frac{1}{v} \int_{\Gamma_0} N \cdot P_{\text{Micro}}X d\Gamma_0 \tag{28}$$

where n and N are the normal vectors in initial (Γ) and current (Γ_0) RVE boundaries, respectively. The integration in equation (28) can be simplified for prescribed displacement BC and a periodic BC. The consistent tangent stiffness matrix can also be calculated using the RVE equilibrium equation and the relation between averaged RVE stress and its averaged work conjugate quantity. Next, macroscale stress and stiffness matrix calculations are treated separately for the prescribed displacement BC and periodic BC.

4.3 Prescribed displacement BC

The macroscale deformation gradient is used as a mapping tensor on the boundaries of the RVE:

$$x = F_{\text{Macro}} \cdot X \qquad \text{with X on } \Gamma_0 \tag{29}$$

where x refers to deformed and X to undeformed position vectors of the points along the boundary, Γ_0 is the initial undeformed boundary points of the RVE.

For the case of prescribed displacement BC, integration of equation (26) simply leads to the following summation

$$P_{\text{Macro}} = \frac{1}{V_0} \Sigma_i \; p_{(i)} X_{(i)} = \frac{1}{V_0} \Sigma_i \; f_{(i)} X_{(i)} \tag{30}$$

where $f_{(i)}$ are the resulting external forces at the boundary nodes (that are caused by P_{Macro} stress on these nodes). In the preceding equation, $X_{(i)}$ is the position vector of the i-th node along the undeformed boundary Γ_0 of RVE. The macroscale 1st PK stress in equation (30) is a non-symmetric second-order tensor, and in component form for two dimensions it can be written as follows

$$P_{\text{Macro}} = \frac{1}{V_0} \begin{bmatrix} \Sigma_i \; f_{1(i)} X_{1(i)} & \Sigma_i \; f_{1(i)} X_{2(i)} \\ \Sigma_i \; f_{2(i)} X_{1(i)} & \Sigma_i \; f_{2(i)} X_{2(i)} \end{bmatrix} \tag{31}$$

where scalar values $f_{1(i)}$ and $f_{2(i)}$ are resulting external forces in basis direction one and two, respectively. Similarly, scalar values $X_{1(i)}$ and $X_{2(i)}$ refer to the initial position vector in directions one and two. The subscript index (i) indicates the quantity is on the i-th node along the boundary.

Multiscale constitutive relations are required to determine the macroscale consistent stiffness matrix at $t = t + 1$. The components of this matrix are the derivative of the macroscale 1st PK stress with respect to its corresponding work conjugate, macroscale deformation gradient

$$C = \frac{\partial P_{\text{Macro}}}{\partial F_{\text{Macro}}} \tag{32}$$

where C is the fourth-order tensor of the macroscale consistent stiffness moduli. In the incremental iterative solver, the above derivative can be rewritten in terms of iterative variations of 1st PK stress and deformation gradients

$$C = \frac{\delta P_{\text{Macro}}}{\delta F_{\text{Macro}}} \tag{33}$$

Equation (31) can be used to find variations of 1st PK stress

$$\delta P_{Macro} = \frac{1}{V_0} \begin{bmatrix} \sum_i \delta f_{1(i)} X_{1(i)} & \sum_i \delta f_{1(i)} X_{2(i)} \\ \sum_i \delta f_{2(i)} X_{1(i)} & \sum_i \delta f_{2(i)} X_{2(i)} \end{bmatrix} \tag{34}$$

To find variations in external forces $\delta f_{(i)}$ one can use the RVE equilibrium equation written in matrix notation as

$$\begin{bmatrix} K_{dd} & K_{db} \\ K_{bd} & K_{bb} \end{bmatrix} \begin{bmatrix} \delta u_d \\ \delta u_b \end{bmatrix} = \begin{bmatrix} \delta f_a \\ \delta f_b \end{bmatrix} \tag{35}$$

where subscript b refers to degrees of freedom of the boundary points and subscript d refers to degrees of freedom of internal points in the RVE domain. After solution of the RVE model, Equation (35) can be reduced to find the stiffness matrix that corresponds only to the prescribed degrees of freedom on the boundary points

$$[K_{bb} - K_{bd}(K_{dd})^{-1}K_{db}]\delta u_b = \delta f_b \tag{36}$$

where the left hand side term in the bracket is stiffness matrix K_{Micro_b}. Equation (36) may be rewritten in tensor notation.

$$\sum_j [K_{bb} - K_{bd}(K_{dd})^{-1}K_{db}]_{(ij)} \delta u_{(j)} = \delta f_{(i)} \tag{37}$$

Subscripts i and j refer to nodes on the boundary Γ_0 of the RVE where the displacement is prescribed.

Now by substitution of equation (37) into equation (34) the numerator term in (33) is found as

$$\delta P_{Macro} = \frac{1}{V_0} \sum_i \sum_j K_{(ij)Micro\_b} \delta u_{(j)} X_{(i)} \tag{38}$$

On prescribed nodes, the displacement is $u_{(j)} = (F_{Macro} - I) \cdot X_{(j)}$, and their variation is simply $\delta u_{(j)} = X_{(j)} \cdot \delta F_{Macro}$. Hence, equation (38) simplifies to

$$\delta P_{Macro} = \frac{1}{V_0} \sum_i \sum_j X_{(j)} K_{(ij)Micro\_b} X_{(i)} : \delta F_{Macro} \tag{39}$$

Comparing equation (39) and (33), the macroscale consistent tangent stiffness matrix is found from the right hand side of equation (39) to be

$$C = \frac{1}{V_0} \sum_i \sum_j X_{(j)} K_{(ij)Micro\_b} X_{(i)} \tag{40}$$

where each component of the matrix on the right hand side is a [2x2] matrix itself. In other words, $K_{(ij)Micro\_b}$ is the same as a [2x2] matrix found from the global microscale stiffness matrix at rows and columns of degrees of freedom i and j.

4.4 Periodic BC

The macroscale deformation gradient is used as a mapping tensor on opposite boundaries of the RVE such that the undeformed and deformed shapes of opposite edges of the RVE are always the same and the stresses are equal but in an opposite direction.

$$x^+ - x^- = F_{macro} \cdot (X^+ - X^-) \text{ and } P^+ = P^- \tag{41}$$

where $p = N \cdot P_{micro}$ is the traction related to the 1st PK stress tensor and normal N with reference to the initial configuration of the RVE. Some writers have modified the periodic BC of equations (41) to simple relations between top-bottom and right-left edges of an initially square shape two-dimensional RVE, shown in Figure 1 (Kouznetsova et al. 2001; Smit et al. 1999).

The choice of periodic boundary conditions has been proven to give a more reasonable estimation of the averaged properties, even when the microstructure is not regularly periodic (Kouznetsova et al. 2001; Smit et al. 1999; Van der Sluis et al. 2001). The periodic boundary condition implies that kinematic (or essential) boundary conditions are periodic and natural boundary conditions are antiperiodic. The latter is to maintain the stress continuity across the boundaries. The spatial periodicity of N is enforced by the following kinematic boundary constraints

$$X_{Top}(\xi_{34}) + X_1 - X_4 - X_{Bottom}(\xi_{12}) = 0 \text{ for } \xi_{34} = \xi_{12}$$

$$X_{Right}(\xi_{23}) + X_1 - X_2 - X_{Left}(\xi_{14}) = 0 \text{ for } \xi_{23} = \xi_{41}$$
(42)

and natural boundary constraints

$$\sigma \cdot n_{Top}(\xi_{34}) = -\sigma \cdot n_{Bottom}(\xi_{12}) \quad \text{for} \quad \xi_{34} = \xi_{12}$$

$$\sigma \cdot n_{Right}(\xi_{34}) = -\sigma \cdot n_{Left}(\xi_{12}) \quad \text{for} \quad \xi_{23} = \xi_{41}$$
(43)

where on the ij-th boundary, ξ_{ij} and η_{ij} are local coordinates of nodes on the top-bottom and left-right edges, respectively, as shown in Figure 1. Assuming that point C1 is located on the origin of the rectangular Cartesian coordinate system of (ξ, η), vertex 1 is fully fixed to eliminate rigid body rotation of the RVE. The kinematic boundary condition (or prescribed displacement due to the macroscale deformation gradient) acts only on the three corner nodes C1, C2 and C4.

$$u_i = (F_{macro} - I)X_i \text{ for } i = C1, C2 \text{ and } C4$$
(44)

Using equations (41)-(44) one can fully define the boundary value problem of the RVE model. Because in a displacement-based code, stress periodicity will be approximately satisfied due to the displacement periodicity induced by constraints of equation (42), the periodicity condition may be recast in terms of displacement constraints only.

$$u_{Right} = u_{Left} - u_2 + u_1$$

$$u_{Top} = u_{Bottom} - u_1 + u_4$$
(45)

Equations (44) and (45) provide a link between degrees of freedom of top-bottom and right-left boundaries of the RVE in the microscale global stiffness matrix.

For the periodic boundary condition, the calculation of the macroscale stress follows that of the prescribed displacement case and is found by equation (34). The only difference is that subscript i refer to the three prescribed corner nodes of C1, C2 and C4.

Similarly, the macroscale consistent tangent moduli is found by equation (40) whereas again subscripts i and j correspond only to corner nodes 1, 2 and 4. Furthermore, as shown in Figure 2, the global stiffness matrix for the microscale is partitioned to points on corners (indicated by subscript c), points on positive (subscript p) and negative (subscript n) edges, as well as points inside the domain of the RVE model (subscript d)

$$\begin{bmatrix} K_{dd} & K_{dp} & K_{dn} & K_{dc} \\ K_{pd} & K_{pp} & K_{pn} & K_{pc} \\ K_{nd} & K_{np} & K_{nn} & K_{nc} \\ K_{cd} & K_{cp} & K_{cn} & K_{cc} \end{bmatrix} \begin{bmatrix} \delta u_d \\ \delta u_p \\ \delta u_n \\ \delta u_c \end{bmatrix} = \begin{bmatrix} \delta f_d \\ \delta f_p \\ \delta f_n \\ \delta f_c \end{bmatrix} \tag{46}$$

where the external force vector of points inside the RVE domain satisfies $\delta f_d = 0$.

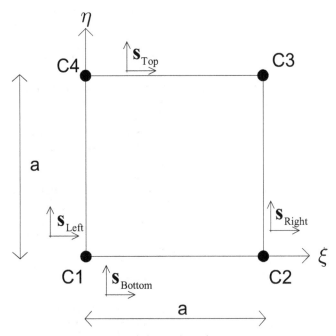

Fig. 1. Typical square shape RVE in undeformed configuration

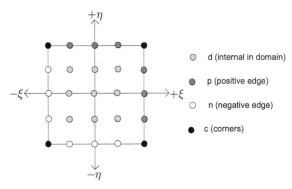

Fig. 2. Partitioning of global stiffness matrix of microscale with reference to position of material points in an RVE with periodic boundary conditions

The periodic boundary condition then implies that on the microscale $\delta f_p = -\delta f_n$, $\delta u_p = \delta u_n$ and $\delta u_c = 0$. The rest of the procedure to calculate the consistent tangent stiffness matrix for the macroscale material point is similar to calculations described in previous section for the prescribed BC.

5. Numerical models

The multiscale PIC method developed in the previous sections was applied to a series of case studies to assess its efficiency and accuracy compared to traditional fluid mechanics or solid mechanics modeling and simulation methods. Two examples are presented that employ the PIC method, without any multiscale effect, to showcase the differences between conventional FE approach and PIC method. These examples are followed by some more examples of metal forming processes modeled using the multiscale PIC method described in previous section.

5.1 Shear banding of viscoelastic plastic material under incompressible viscous flows
This example is a 2D numerical experiment of a strain-softening viscoelastic-plastic material undergoing simple shear in a periodic domain. We anticipate the development of strong shear banding in this situation. Our interest lies in understanding 1) the initial orientation of the shear bands as they develop from a random distribution of initial softening, 2) the high-strain configuration of the shear bands, and how deformation is accommodated after the shear bands become fully softened. This experiment is similar to the ones described in Lemiale et al. (2008); here we use simple-shear boundary conditions and very high resolution to study several generations of shear bands. The sample material is confined between two plates of viscous material with "teeth" which prevent shear-banding from simply following the material boundaries. A small plug of non-softening material is applied at each end of the domain to minimize the effect of shear bands communicating across the wrap-around boundary. The example shown in Figures 3 and 4 are of a material with 20% cohesion softening and a friction coefficient of 0.3 (which remains constant through the experiment). The material properties of the sample are: viscosity = 100, elastic shear modulus = 10^6, cohesion = 50. This ensures that the material begins to yield in the first few steps after the experiment is started.

A large number of shear bands initiate from the randomly distributed weak seeds in two conjugate directions in equal proportions. These two orientations are: 1) approximately 5 degrees to the maximum velocity gradient (horizontal) and 2) the conjugate to this direction which lies at 5 degrees to the vertical. These orientations are highlighted in Figure 3a which is a snapshot taken at a strain of 0.6%.

As the strain increases to 3% (Figure 3.b), the initial shear bands coalesce into structures with a scale comparable to the sample width. The orientation of the largest shear bands in the interior of the sample is predominantly at 10 degrees to the horizontal. This angle creates a secondary shear orientation, and further, smaller shear bands are seen branching from the major shear bands.

At total strain between 100% and 130%, the shear bands have begun to reorganize again into structures sub-parallel to the maximum velocity gradient. Softening has concentrated the deformation into very narrow shear bands which can be seen in Figure 3c and 3d. Although the total strain is of order unity, the maximum strain in the shear bands is higher by nearly two orders of magnitude. Further deformation of the sample produces almost no deformation in the bulk of the material; all subsequent deformation occurs on the shear bands (Figure 3c evolving to Figure 3d).

The progression of the experiment from onset to a total shear across the sample of 130% is shown in Figure 3. The plot is of plastic strain overlain with material markers. Figures 4a and 4b are the full-size experiments from which Figure 3a and 3b were taken. Figures 4(e,f) are the full-size images of Figure 3(c,d).

Although this experiment represents solid deformation, the extremely high-strains present in the shear bands warrant the formulation of the problem using a fluid approach.

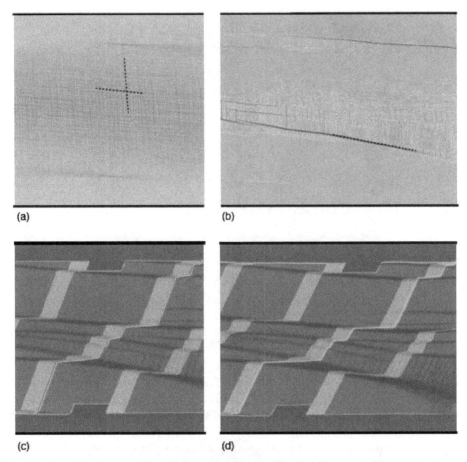

(a)   (b)

(c)   (d)

Fig. 3. A viscoplastic material with a strain-softening Drucker-Prager yield criterion is subjected to a simple shear boundary condition which gives a velocity gradient of 1 across the sample. The resolution is 1024x256 elements in domain of 8.0 x 2.0 in size. The strain rate is shown in (a,b) for an applied strain of 0.6% and 3% respectively. The scale is logarithmic varying from light blue (strain rates < 0.01) to dark red (strain rates > 5.0). The average shear strain rate applied to the experiment is 1.0. Shear bands are fully developed at 1-2% strain. At high strains of 100% and 130%, shown in (c,d), almost all deformation occurs in the shear bands. In these examples we have included light-coloured stripes, initially vertical, which mark the strain. The shear bands are visualized by applying dark colouring to material which reaches the maximum plastic strain of 1.0.

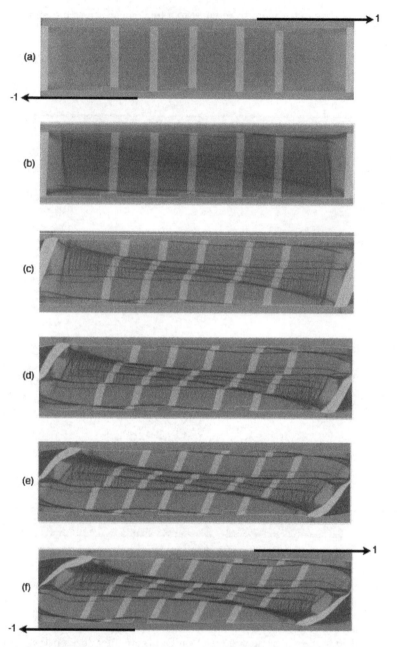

Fig. 4. The evolution of the shear experiment from 0.6% applied strain (a) through to 130% strain (f). The intermediate values of strain are 3% (b), 33% (c), 66% (d) and 100% (e). The resolution is 1024x256 elements in domain of 8.0 x 2.0 in size; strain markers and shear bands are visualized as in Fig. 3c,d.

5.2 Viscoelasticity in subduction models

The model in which we demonstrate the importance of viscoelastic effects in subduction is one in which an isolated oceanic plate founders into the mantle. Although this model is not a faithful representation of a subduction zone, it is well understood, and different regimes of behavior have been observed and interpreted in a geological context for purely viscous plates (OzBench et al. 2008). The model has a viscoelastic inner core with viscoplastic outer layers. The lithosphere undergoes significant strain as it traverses the fluid (mantle) layer.

Figure 5 shows subduction models in which μ ranges across 3 orders of magnitude with a constant value of the viscosity contrast between the core of the slab and the background mantle material of $\Delta\eta = 2 \times 10^4$, resulting in relaxation times from 20,000 years through to 2 million years. Contour lines of viscosity are included to differentiate between the upper mantle, outer plastic lithosphere and slab core.

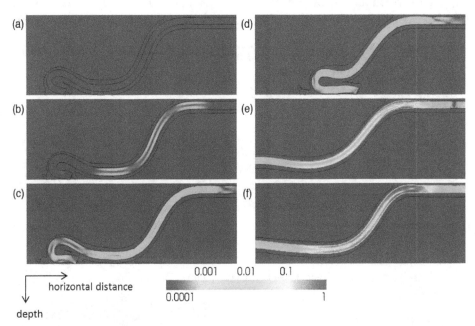

Fig. 5. Weissenberg number for free slab models at steady state with increasing elasticity. The models represent the outer 670km of the Earth. The models are run in a domain of aspect ratio 6; the images are then translated to show the hinge at the same position in the diagram for each of the cases. This steady state snapshot is taken when the subduction rate reaches a constant value and with the slab fully supported by the lower mantle. (a) Viscous only core $\Delta\eta = 2 \times 10^4$. Viscoelastic core models have $\Delta\eta = 2 \times 10^4$ with a scaled observation time (Δt_e) of 2×10^4 yrs for all models and elastic modulus varying as (b) $\mu = 2 \times 10^{11}$ Pa ($\alpha = 21$ Myr) (c) $\mu = 8 \times 10^{10}$ Pa ($\alpha = 108$ Myr) (d) $\mu = 4 \times 10^{10}$ Pa ($\alpha = 194$ Myr) (e) $\mu = 8 \times 10^9$ Pa ($\alpha = 1080$ Myr) (f) $\mu = 4 \times 10^9$ Pa ($\alpha = 2160$ Myr)

The coloring indicates the Weissenberg number: a measure of the relaxation time to the local characteristic time of the system (here defined by the strain rate). In regions where this value is close to unity, the role of elasticity dominates the observed deformation. The increase in

Weissenberg number highlights the change in slab morphology during steady state subduction as elasticity is increased with a fold and retreat mode observed for the viscous core model (a) and models (b-d). As μ decreases, the elastic stresses increase, producing in a strongly retreating lithosphere for low values of μ.

In Figure 6, we show the stress orientations associated with the viscosity-dominated and elasticity-dominated models. The stress distribution and orientation each influence seismicity and focal mechanisms of earthquakes. Although the near-surface morphology of the lithosphere is quite similar in each case, the distinct patterns of stress distribution and the difference in stress orientation during bending indicate very different balances of forces. Viscoelastic effects are clearly important in developing models of the lithospheric deformation at subduction zones. Particle based methods such as ours allow the modeling not only of the viscoelastic slab, but the continuous transition through the thermal boundary layer, to the surrounding viscous mantle. This makes it possible to study, for example, the interaction of multiple slabs in close proximity, or slabs tearing during continental collision.

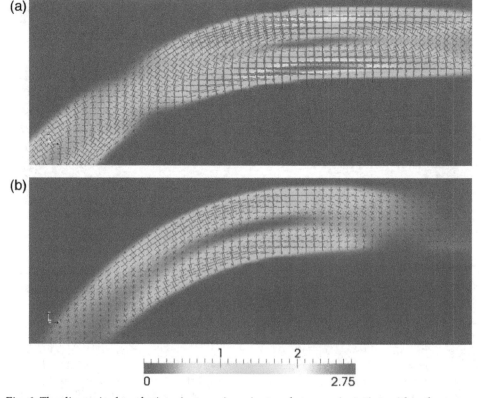

Fig. 6. The dimensionless deviatoric stress invariant and stress orientation within the core showing extension (red) and compression (blue) axes at the steady state time step shown in Figure 5 for a relaxation time of (a) 21 k yr – viscosity dominated and (b) 2,160 k yr – elasticity playing a dominant role in the hinge region. The eigenvectors are plotted using the same scale for both (a) and (b). The reference stress value is 48 MPa.

5.3 Multiscale modeling of AHSS

Advanced High Strength Steels such as Dual Phase (DP) and Transformation Induced Plasticity (TRIP) steels have a multiphase microstructure that strongly affects their forming behavior and mechanical properties. DP steel's microstructure consists of hard martensite inclusions in a ferrite matrix. TRIP steel has a similar microstructure with addition of retained austenite which then potentially transforms into martensite during deformation. The result of this transformation is better combination of ductility and strength. The multiphase nature of the DP and TRIP steels' microstructure is the focus of modeling with multiscale PIC method as well as with conventional FE method (Asgari et al. 2008).

One of the difficulties with FEM micromechanical models is the requirement for mesh refinement. Our multiscale PIC method benefits from the h-type of refinement similar to the one traditionally used in FEM. Using this approach, it is possible to increase the number of background cells as if the cells were elements in a FE model. This enrichment in the multiscale PIC model is schematically shown in Figure 7 for a three phase (TRIP steel) microscale model. Accordingly, convergence tests were used to certify that the solution was not altered by the increasing number of background cells.

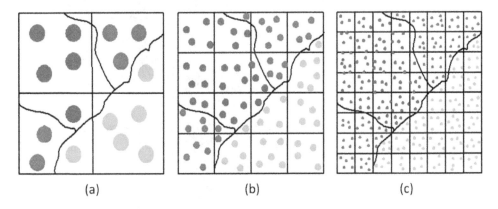

      (a)   (b)   (c)

Fig. 7. The multiscale PIC enrichment similar to the idea of h-refinement in FEM

With the aid of background cell enrichment, an important feature of the multiscale PIC method is also visualised in Figure 7. This feature is demonstrated in the capability of the multiscale PIC method to represent two or three (or even more) material properties within a single background cell; for example in Figure 7a, the lower left hand side cell contains three different material points. Such representation in FE method needs at least three elements with three different material properties. This advantage of PIC method was carried over into the refined models as shown in Figure 7(b,c). In these cases, the phase interfaces and background cell boundaries never disturbed each other and are independent.

The simplified unit cell configuration was used in microscale analyses of DP and TRIP steels, as shown in Figure 8. In addition to simplified unit cell representation, it is possible to use realistic microstructure of these steels in the simulations. For further details on strength to realistic microstructure see Asgari et al. (Asgari et al. 2009). The two phases in the DP microscale models and the three phases in the TRIP microscale models obeyed the J2 theory of deformation. For these constituent phases being isotropic elastoplastic, only two elastic constants were necessary to describe the elastic behavior, and Swift hardening was assumed

$$\sigma(\varepsilon^{pl}) = \sigma_{Y0}(1 + H\varepsilon^{pl})^n \tag{47}$$

Where ε^{pl} is the accumulated plastic strain. σ_{Y0} was considered to be 500, 780 and 2550 MPa for ferrite, austenite and martensite, respectively. The hardening factor and hardening exponent (H, n) were taken to be (93 MPa, 0.21) for ferrite, (50 MPa, 0.22) for austenite and (800 MPa, 0.2) for martensite. These material properties can be obtained by a combination of neutron diffraction, nano-indentation and microstructural imaging described by Jacques et al. (Jacques et al. 2007).

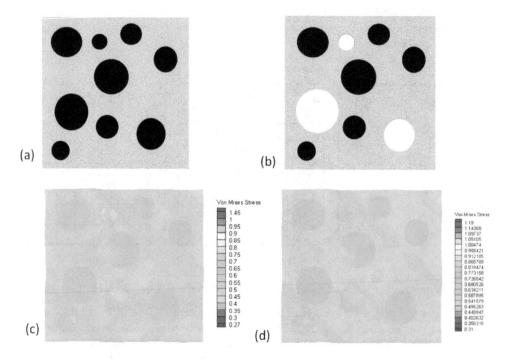

Fig. 8. Geometrical representation of the simplified unit cells for a) DP steel with 76.76% ferrite (grey) and 23.24% martensite (black) and b) TRIP steel with 77.12% ferrite (Gray), 14.75% austenite (black) and 8.13% martensite (white); and the von Mises stress distribution for c) DP steel and d) TRIP steel

The PIC method was quantitatively stiffer than FEM at the peak stress, although in some local regions around corners of the microscale unit cell boundary, FEM occasionally showed more stiffness than PIC. The effective plastic stress and strain predicted from the simplified unit cells were calculated using the volume averaging homogenization technique. These are plotted in Figures 9(a,b) for the DP and TRIP steel, respectively. The effective stress predicted from simplified regular unit cells showed up to 10% error using the FE method and up to 6% error using the Multiscale PIC method, which does not prove to be significantly more accurate than FE method.

An interesting observation made from the macroscale effective stress of simplified unit cells was the prediction of the yield point using FE and multiscale PIC method. In the case of the DP steel, both methods were able to predict the yield stress quite accurately.

However, in the case of the TRIP steel, the multiscale PIC method produced more error compared to the FE results. The increased error occurs because in multiscale PIC method the location of integration points are not at the optimal locations as the FE integration points are. Therefore, there might be some loss of accuracy due to integrations performed in this method. This loss of accuracy balances out with the improved accuracy obtained from the smoother interpolation of the field variables (especially on and around the phase interfaces) after a certain strain values. Therefore, the error in the FE method continued increasing while that of the multiscale PIC method reached a plateau, and might have even decreased if deformed towards much larger strains.

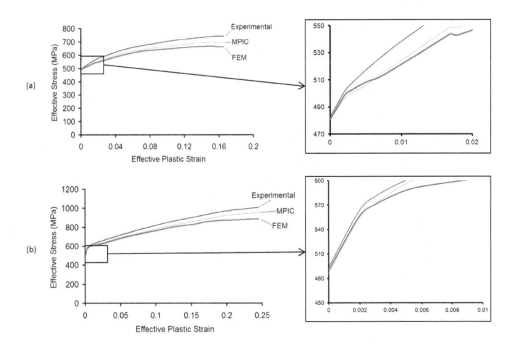

Fig. 9. The effective plastic macro stress from simplified regular unit cell models of a) DP steel and b) TRIP steel, showing comparison between FE method (FEM), multiscale PIC (MPIC) and experimental stress strain data.

6. Conclusion

The examples of large strain deformation of viscoelastic-plastic material and tectonic plates shows that PIC method is very suitable for solid mechanics problems where large deformations are encountered. However, the last example showed that the multiscale PIC method does not have any significant advantage over FE method in small strains for solid mechanic formulation. In such cases, the maturity of the FE method dominates the minimal accuracy benefits that could be obtained by using material particles instead of integration points.

7. Acknowledgement

Authors would like to acknowledge funding and support from Institute for Technology, Research and Innovation (ITRI) and Victorian Partnership for Advanced Computing (VPAC). Code development was partly supported by AuScope, a capability of the National Collaborative Research Infrastructure Strategy (NCRIS).

8. References

Asgari, S. A., Hodgson, P. D., Lemiale, V., Yang, C. & Rolfe, B. F. (2008). Multiscale Particle-In-Cell modelling for Advanced High Strength Steels. *Advanced Materials Research*, Vol. 32, No. 1, pp. 285-288

Asgari, S. A., Hodgson, P. D. & Rolfe, B. F. (2009). Modelling of Advanced High Strength Steels with the realistic microstructure-strength relationships. *Computational Materials Science*, Vol. 45, No. 4, pp. 860-886

Belytschko, T., Krongauz, Y., Organ, D., Fleming, M. C. & Krysl, P. (1996). Meshless methods: An overview and recent developments. *Comput. Methods Appl. Mech. Engrg.*, Vol. 139, No., pp. 3-47

Belytschko, T., Liu, W. K. & Moran, B. (2001). Nonlinear FEs for continua and structures John Wiley & Sons Inc., West Sussex

Benson, D. (1992). Computational methods in Lagrangian and Eulerian hydrocodes. *Comput. Methods Appl. Mech. Engrg.*, Vol. 99, No., pp. 235-394

Chen, J. S., Pan, C., Roque, C. M. O. L. & Wang, H. P. (1998). A Lagrangian reproducing kernel particle method for metal forming analysis. *Computational Mechanics*, Vol. 22, No., pp. 289-307

Guilkey, J. E. & Weiss, J. A. (2003). Implicit time integration for the material point method: Quantitative and algorithmic comparisons with the finite element method. *International Journal for Numerical Methods in Engineering*, Vol. 57, No., pp. 1323-1338

Jacques, P. J., Furnemont, Q., Lani, F., Pardoen, T. & Delannay, F. (2007). Multiscale mechanics of TRIP-assisted multiphase steels: I. Characterization and mechanical testing. *Acta Materialia*, Vol. 55, No., pp. 3681-3693

Karato, S. (1993). Rheology of the upper mantle - a synthesis. *Science*, Vol. 260, No., pp. 771-778

Kouznetsova, V. (2002). *Computational homogenization for the multi-scale analysis of multi-phase materials*. Department of Mechanical Engineering. pp. 120, Eindhoven University of Technology, Eindhoven

Kouznetsova, V., Brekelmans, W. A. M. & Baaijens, F. P. T. (2001). An approach to micro-macro modeling of heterogeneous materials. *Computational Mechanics*, Vol. 27, No., pp. 37-48

Kouznetsova, V., Geers, M. G. D. & Brekelmans, W. A. M. (2004). Size of a Representative Volume Element in a second-order computational homogenization framework. *International Journal for Multiscale Computational Engineering*, Vol. 2, No. 4, pp. 575-598

Lemiale, V., Muhlhaus, H. B., Moresi, L. & Stafford, J. (2008). Shear banding analysis of plastic models formulated for incompressible viscous flows. *Physics of the Earth and Planetary Interiors*, Vol. 171, No. 1-4, pp. 177-186

Liu, W. K., Karpov, E. G., Zhang, S. & Park, H. S. (2004). An introduction to computational nanomechanics and materials. *Comput. Methods Appl. Mech. Engrg.*, Vol. 193, No., pp. 1529-1578

Moresi, L., Dufour, F. & Muhlhaus, H. B. (2003). A lagrangian integration point finite element method for large deformation modeling of viscoelastic geomaterials. *J. Comp. Physics*, Vol. 184, No., pp. 476-497

OzBench, M., Regenauer-Lieb, K., Stegman, D. R., Morra, G., Farrington, R., Hale, A., May, D. A., Freeman, J., Bourgouin, L., Muhlhaus, H. B. & Moresi, L. (2008). A model comparison study of large-scale mantle-lithosphere dynamics driven by subduction. *Physics of the Earth and Planetary Interiors*, Vol. 171, No. 1-4, pp. 224-234

Smit, R. J. M., Brekelmans, W. A. M. & Meijer, H. E. H. (1998). Prediction of the mechanical behavior of nonlinear heterogeneous systems by multi-level finite element modeling. *Comput. Methods Appl. Mech. Engrg.*, Vol. 155, No., pp. 191-192

Smit, R. J. M., Brekelmans, W. A. M. & Meijer, H. E. H. (1999). Prediction of the large-strain mechanical response of heterogeneous polymer systems: local and global deformation behaviour of a representative volume element of voided polycarbonate. *Journal of the Mechanics and Physics of Solids*, Vol. 47, No., pp. 201-221

Sulsky, D. & Schreyer, H. L. (1996). Axisymmetric form of the material point method with applications to upsetting and Taylor impact problems. *Comput. Methods Appl. Mech. Engrg.*, Vol. 139, No., pp. 409-429

Sulsky, D., Zhou, S. J. & Schreyer, H. L. (1995). Application of a particle-in-cell method to solid mechanics. *Computer Physics Communications*, Vol. 87, No., pp. 236-252

Van der Sluis, O., Schreurs, P. J. G. & Meijer, H. E. H. (2001). Homogenisation of structured elastoviscoplastic solids at finite strains. *Mechanics of Materials*, Vol. 33, No., pp. 499-522

Watts, A. B., Bodine, J. H. & Ribe, N. M. (1980). Observations of flexure and the geological evolution of the pacific basin. *Nature*, Vol. 283, No., pp. 532-537

Permissions

The contributors of this book come from diverse backgrounds, making this book a truly international effort. This book will bring forth new frontiers with its revolutionizing research information and detailed analysis of the nascent developments around the world.

We would like to thank Prof. Steven A. Jones, for lending his expertise to make the book truly unique. He has played a crucial role in the development of this book. Without his invaluable contribution this book wouldn't have been possible. He has made vital efforts to compile up to date information on the varied aspects of this subject to make this book a valuable addition to the collection of many professionals and students.

This book was conceptualized with the vision of imparting up-to-date information and advanced data in this field. To ensure the same, a matchless editorial board was set up. Every individual on the board went through rigorous rounds of assessment to prove their worth. After which they invested a large part of their time researching and compiling the most relevant data for our readers. Conferences and sessions were held from time to time between the editorial board and the contributing authors to present the data in the most comprehensible form. The editorial team has worked tirelessly to provide valuable and valid information to help people across the globe.

Every chapter published in this book has been scrutinized by our experts. Their significance has been extensively debated. The topics covered herein carry significant findings which will fuel the growth of the discipline. They may even be implemented as practical applications or may be referred to as a beginning point for another development. Chapters in this book were first published by InTech; hereby published with permission under the Creative Commons Attribution License or equivalent.

The editorial board has been involved in producing this book since its inception. They have spent rigorous hours researching and exploring the diverse topics which have resulted in the successful publishing of this book. They have passed on their knowledge of decades through this book. To expedite this challenging task, the publisher supported the team at every step. A small team of assistant editors was also appointed to further simplify the editing procedure and attain best results for the readers.

Our editorial team has been hand-picked from every corner of the world. Their multi-ethnicity adds dynamic inputs to the discussions which result in innovative outcomes. These outcomes are then further discussed with the researchers and contributors who give their valuable feedback and opinion regarding the same. The feedback is then collaborated with the researches and they are edited in a comprehensive manner to aid the understanding of the subject.

Apart from the editorial board, the designing team has also invested a significant amount of their time in understanding the subject and creating the most relevant covers. They scrutinized every image to scout for the most suitable representation of the subject and create an appropriate cover for the book.

The publishing team has been involved in this book since its early stages. They were actively engaged in every process, be it collecting the data, connecting with the contributors or procuring relevant information. The team has been an ardent support to the editorial, designing and production team. Their endless efforts to recruit the best for this project, has resulted in the accomplishment of this book. They are a veteran in the field of academics and their pool of knowledge is as vast as their experience in printing. Their expertise and guidance has proved useful at every step. Their uncompromising quality standards have made this book an exceptional effort. Their encouragement from time to time has been an inspiration for everyone.

The publisher and the editorial board hope that this book will prove to be a valuable piece of knowledge for researchers, students, practitioners and scholars across the globe.

List of Contributors

L. Martínez-Suástegui
ESIME Azcapotzalco, Instituto Politécnico Nacional, Colonia Santa Catarina, Delegación Azcapotzalco, México, Distrito Federal, Mexico

Ye. Belyayev
Al-Farabi Kazakh National University, Kazakhstan

A. Naimanova
Institute of Mathematics, Ministry of Education and Science, Kazakhstan

Norhashidah Hj Mohd Ali and Foo Kai Pin
Universiti Sains Malaysia, Malaysia

Sanjay Mishra
School of Engineering Systems, Queensland University of Technology, Brisbane, Australia

Christopher Depcik and Sudarshan Loya
University of Kansas, Department of Mechanical Engineering, USA

C.H. Huang
U.S. Department of the Interior, Bureau of Ocean Energy Management, Regulation, and Enforcement, USA

Leopold Vrankar and Franc Runovc
Faculty of Natural Sciences and Engineering, Slovenia

Goran Turk
Faculty of Civil and Geodetic Engineering, University of Ljubljana, Slovenia

Yoshifumi Ogami
Ritsumeikan University, Japan

Maatoug Hassine
ESSTH, Sousse University, Tunisia

Alireza Asgari
School of Engineering, Deakin University, Australia

Louis Moresi
School of Mathematical Sciences, Monash University, Australia